UMAP

Modules

Tools for Teaching 1997

published by

The Consortium for Mathematics
and Its Applications, Inc.
Suite 210
57 Bedford Street
Lexington, MA 02173–4496

edited by

Paul J. Campbell
Campus Box 194
Beloit College
700 College Street
Beloit, WI 53511–5595
campbell@beloit.edu

Recommendations expressed are those of the authors and do not necessarily reflect the views of the copyright holder.

ISBN 0-912843-46-2
Printed in U.S.A.

Table of Contents

COMAP

Introduction

The instructional Modules in this volume were developed by the Undergraduate Mathematics and Its Applications (UMAP) Project. Project UMAP develops and disseminates instructional modules and expository monographs in mathematical modeling and applications of the mathematical sciences, for undergraduate students and their instructors.

UMAP Modules are self-contained (except for stated prerequisites) lesson-length instructional units. From them, undergraduate students learn professional applications of the mathematical sciences. UMAP Modules feature different levels of mathematics, as well as various fields of application, including biostatistics, economics, government, earth science, computer science, and psychology. The Modules are written and reviewed by instructors in colleges and high schools throughout the United States and abroad, as well as by professionals in applied fields.

UMAP was originally funded by grants from the National Science Foundation to the Education Development Center, Inc. (1976–1983) and to the Consortium for Mathematics and Its Applications (COMAP) (1983–1985). In order to capture the momentum and success beyond the period of federal funding, we established COMAP as a nonprofit educational organization. COMAP is committed to the improvement of mathematics education, to the continuing development and dissemination of instructional materials, and to fostering and enlarging the network of people involved in the development and use of materials. In addition to involvement at the college level through UMAP, COMAP is engaged in science and mathematics education in elementary and secondary schools, teacher training, continuing education, and industrial and government training programs.

In addition to this annual collection of UMAP Modules, other college-level materials distributed by COMAP include individual Modules (more than 500), *The UMAP Journal*, and UMAP expository monographs. Thousands of instructors and students have shared their reactions to the use of these instructional materials in the classroom, and comments and suggestions for changes are incorporated as part of the development and improvement of materials.

This collection of Modules represents the spirit and ability of scores of volunteer authors, reviewers, and field-testers (both instructors and students). The substance and momentum of the UMAP Project comes from the thousands of individuals involved in the development and use of UMA instructional materials. COMAP is very interested in receiving information on the use of Modules in various settings. We invite you to call or write for a catalog of available materials, and to contact us with your ideas and reactions.

Sol Garfunkel, COMAP Director
Paul J. Campbell, Editor

Recruitment, editing, and selection UMAP Modules is done by the board of editors of *The UMAP Journal*, who are appointed by the editor-in-chief in consultation with the presidents of the cooperating organizations

Mathematical Association of America (MAA),
Society for Industrial and Applied Mathematics (SIAM),
National Council of Teachers of Mathematics (NCTM),
American Mathematical Association of Two-Year Colleges (AMATYC),
Institute for Operations Research and the Management Sciences (INFORMS), and
American Statistical Association (ASA).

In 1997 the editor and associate editors were:

Manuscripts are read double-blind by two or more referees, including an associate editor. Guidelines are published in *The UMAP Journal*.

UMAP

Modules in Undergraduate Mathematics and Its Applications

Published in cooperation with

The Society for Industrial and Applied Mathematics,

The Mathematical Association of America,

The National Council of Teachers of Mathematics,

The American Mathematical Association of Two-Year Colleges,

The Institute for Operations Research and the Management Sciences, and

The American Statistical Association.

Module 743

History and Uses of Complex Numbers

Yves Nievergelt

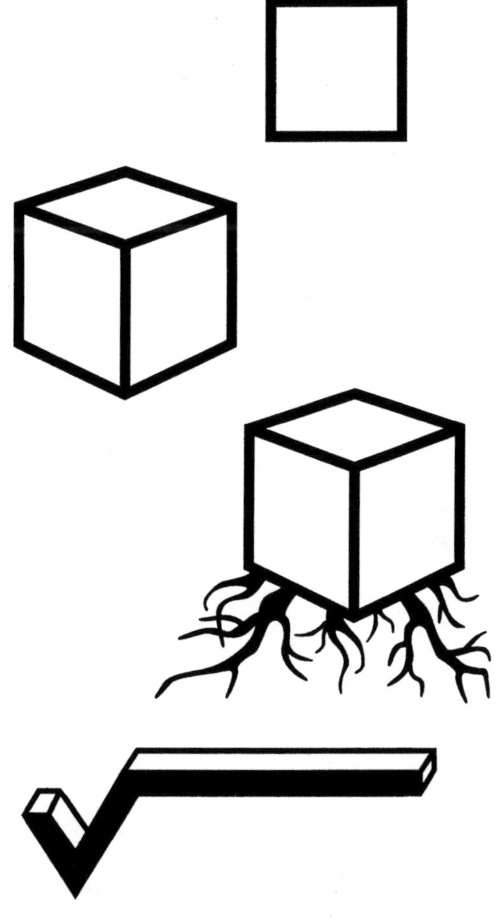

Applications of Algebra, Calculus, Geometry, and Trigonometry to Engineering and Geodesy

COMAP, Inc., Suite 210, 57 Bedford Street, Lexington, MA 02173 (781) 862–7878

INTERMODULAR DESCRIPTION SHEET:	UMAP Unit 743
TITLE:	History and Uses of Complex Numbers
AUTHORS:	Yves Nievergelt Dept. of Mathematics, MS 32 Eastern Washington University 526 5th Street Cheney, WA 99004–2415 `ynievergelt@ewu.edu`
MATHEMATICAL FIELD:	Algebra, calculus, geometry, and trigonometry
APPLICATION FIELD:	Engineering and geodesy
TARGET AUDIENCE:	College students without prior study of complex numbers but enrolled in courses that require complex numbers, for instance, linear algebra, or numerical analysis.
ABSTRACT:	We present the history, the algebra, the geometry, and a few uses of complex numbers, including complex square roots and complex quadratic equations. We then explain the algebraic solutions of cubic and quartic equations; though not using material beyond high school, such explanations may require mathematical perseverance and tenacity. Finally, we demonstrate how complex numbers arise in geodesy.
PREREQUISITES:	A command of high-school geometry and algebra at the level of the derivation of the quadratic formula, including the concepts of associativity, commutativity, and distributivity, and the canonical dot product of vectors in the plane. Some computations of the solutions of the cubic and quartic equations also employ elementary trigonometric functions and identities; some isolated exercises involve calculus or matrix algebra.
RELATED UNITS:	Unit 722: *Using Real Quaternions to Represent Rotations in Three Dimensions*, by Bryant A. Julstrom. Reprinted in *The UMAP Journal* 13 (2) (1992): 121–148 and in *UMAP Modules: Tools for Teaching 1992*, edited by Paul J. Campbell, 1–34. Lexington, MA: COMAP, 1993. Unit 640: *Internal Rates of Return*, by Hiram Paley, Peter F. Colwell, and Roger E. Cannaday. Reprinted in *UMAP Modules: Tools for Teaching 1983*, 493–548. Lexington, MA: COMAP, 1984. Unit 652: *Spacecraft Attitude, Rotations and Quaternions*, by Dennis Pence. Reprinted in *The UMAP Journal* 5 (2) (1984): 215–250 and in *UMAP Modules: Tools for Teaching 1984*, edited by Paul J. Campbell, 129–172. Lexington, MA: COMAP, 1985.

COMAP, Inc., Suite 210, 57 Bedford Street, Lexington, MA 02173
(800) 77-COMAP = (800) 772-6627, or (781) 862-7878; `http://www.comap.com`

History and Uses of Complex Numbers

Yves Nievergelt
Dept. of Mathematics, MS 32
Eastern Washington University
526 5th Street
Cheney, WA 99004–2415
`ynievergelt@ewu.edu`

Table of Contents

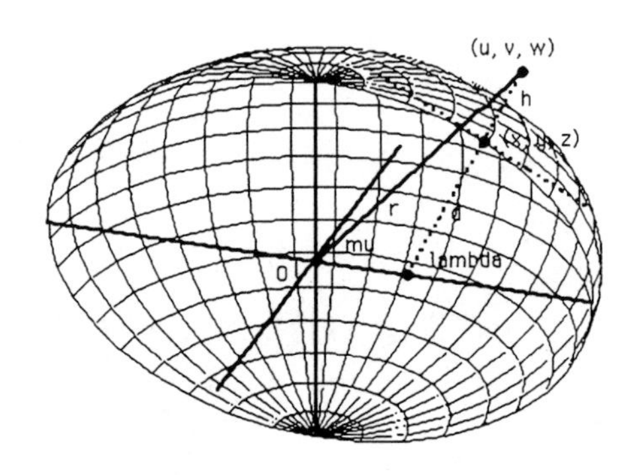

Modules and Monographs in Undergraduate Mathematics and its Applications (UMAP) Project

The goal of UMAP is to develop, through a community of users and developers, a system of instructional modules in undergraduate mathematics and its applications, to be used to supplement existing courses and from which complete courses may eventually be built.

The Project was guided by a National Advisory Board of mathematicians, scientists, and educators. UMAP was funded by a grant from the National Science Foundation and now is supported by the Consortium for Mathematics and Its Applications (COMAP), Inc., a nonprofit corporation engaged in research and development in mathematics education.

Paul J. Campbell	Editor
Solomon Garfunkel	Executive Director, COMAP

1. Introduction

This Module aims to supplement the preparation of university undergraduates without prior study of complex numbers but enrolled in variants of such courses as linear algebra, numerical analysis, or complex analysis, which may require a prior working knowledge of complex numbers. In such courses, many students have a solid command of such prerequisites as calculus, logic, or topology, but they lack the basic working knowledge of complex numbers necessary to study the elements of such topics as Hermitian matrices and the Fast Fourier Transform, or to begin the study of complex analysis. Various versions of this Module have been assigned as independent study to undergraduates during a numerical analysis course at the level of Kincaid and Cheney [1996] and to graduates prior to a complex analysis course at the level of Narasimhan [1985]. The course notes also included applications to chemistry [Macleod 1984], to a few types of financial equations [Paley et al. 1984], and to the design and analysis of electrical circuits [Purcell 1963]. They also included material on linear fractional transformations, but similar material also appears in Hahn [1994] at an elementary level, Spiegel [1964] at an intermediate level, and Schwerdtfeger [1979] at an advanced level. In contrast, elementary introductions to complex numbers, with some indications of their computations and applications, appear scarcer. The material in this Module may also be appropriate for the type of high-school students mentioned in Hahn [1994, viii].

A few "deadly" exercises, however, marked by a dagger (†), first require figuring out the object of the exercise; they then involve a combination of material from outside of complex numbers, great ingenuity, or a lot of calculations, as would problems arising in projects from research or applications.

2. The History of Complex Numbers

The following outline comes largely from Witmer's translation of Cardano [1993], from Struik [1987], and from van der Waerden [1983; 1985].

Complex numbers arose in the Italian mathematician Gerolamo Cardano's book *Ars Magna, sive de regulis algebraicis*, published in Nürnberg in 1545, with a solution to the following problem: Divide 10 into two parts, the product of which equals 40 [van der Waerden 1985, 56 and 177]. In his *Ars Magna*—using previous results from Scipione del Ferro, Niccolò Tartaglia, and Lodovico Ferrari—Cardano also presented algebraic formulae to solve cubic and quartic equations. Such problems originated in antiquity in astronomy, religion, or surveying—for instance, in ancient Greece and India, to appease the gods by constructing altars subject to geometric and algebraic specifications [van der Waerden 1983, 13]. In the Middle Ages, cubic equations occurred with problems about monetary loans—for example, in Dardi of Pisa's *Aliabraa argibra* about 1344 [van der Waerden 1985, 47] and in the painter Piero della Francesca's *Trattato d'Abaco* [van der Waerden 1985, 49]. In the sixteenth century, cubic equations also formed the object of contests for money or goods and services among mathematicians [van der Waerden 1985, 55]. However, Cardano's *Ars Magna* appeared obscure to some, and it omitted cubic equations with three distinct real roots. Algebraic formulae to solve such equations appeared in 1572 in Rafael Bombelli's *l'Algebra* [van der Waerden 1985, 59–61], with $\sqrt{-1}$ denoted by the letter i.

The historical development of complex numbers proved unsatisfactory to some people, because complex numbers had been introduced "temporarily" as a *deus ex machina* without logical relation to previously established mathematics. Even Cardano experienced "mental tortures" in attempting to understand the complex numbers that he discovered for himself [van der Waerden 1985, 56 and 177]. Moreover, complex numbers occurred in academically contrived versions of practical problems. For instance, in sixteenth century Holland, the engineer and mathematician Simon Stevin made the decimal notation and several algebraic symbols popular in Europe but did not use complex numbers, because he felt that they did not help in finding real solutions [van der Waerden 1985, 69]. Nevertheless, in the forms $a + b\sqrt{-1}$ and $a + bi$, complex numbers came into wider use among mathematicians, as René Descartes suggested to "imagine" such numbers in his *Discours de la méthode* in 1637 [van der Waerden 1985, 177]. Later, precise definitions and theoretical foundations evolved in the works of Caspar Wessel (1797), Jean Robert Argand (1806), John Warren (1828), Carl Friedrich Gauss (1831)—who used the phrase "complex numbers"—William Rowan Hamilton (1834)—who adopted the notation (a, b) [Chutsky 1987], [Hamilton 1834]—and Augustin Cauchy (1847), who also considered complex numbers as equivalence classes of polynomials [van der Waerden 1985, 177–178]. Thus, although complex numbers arose from applications and in a somewhat informal manner, precise explanations of complex numbers took shape later, away from applications, in mathematicians' abstractions.

Since their origins in cosmology, finance, and intellectual contests, complex numbers have found other applications to the real world, and they have provided motivation for new mathematics, which in turn applies to other problems. For instance, complex numbers

- ease the computation of trajectories in orbital and space mechanics [Stiefel and Scheifele 1971];
- facilitate the calculation of steady states of the amplitude and phase of alternating electrical currents and potentials, and of the impedance of electrical coils, condensors, and resistors [Hewlett-Packard 1982, 169–171; Purcell 1963, 286–294]; and
- help in solving problems about electric or fluid potentials [Churchill and Brown 1984, Ch. 9], by means of conformal deformations of planar domains [Narasimhan 1985, 148–149].

Also, analogies with complex numbers led Hamilton to the discovery of a yet larger (containing the complex numbers) system of numbers called the *quaternions*, while walking with his wife to a meeting of the Royal Irish Academy in 1843. According to a letter to his son, Hamilton then promptly engraved the main result, $i^2 = j^2 = k^2 = ijk = -1$, onto Brougham Bridge [Halbertstam and Ingram 1967, xv]. Hamilton's quaternions then proved useful in the calculation of rotations in space [Fenwick 1992; Halbertstam and Ingram 1967, 643–644; Julstrom 1992; Schletz 1991], and in controlling the attitude (pitch, roll, and yaw) of spacecraft [Pence 1984; Schletz 1991].

Despite their origins in applied problems, the solution of cubic and quartic equations with complex numbers does not lend itself as well as other methods from numerical analysis to automated computation through the finite arithmetic of digital computers. The principal disadvantage of complex solutions lies in the large number of complex arithmetic operations, which causes difficulties in determining and proving in advance the accuracy of results affected by the computer's rounding inaccuracies. Another disadvantage of complex solutions lies in the necessity of sifting through three or four solutions to identify the one corresponding to the problem under investigation [Miller and Vegh 1993]. Therefore, one of the preferred methods to solve cubic, quartic, or any other equations consists in transforming the problem so that the solution becomes the "fixed point of a contracting map," as described in numerical analysis [Kincaid and Cheney 1996]. In some applications, the fact that the solution arose from a cubic or a quartic equation then becomes irrelevant, and the insistence on using cubic or quartic formulae produces results less accurate than those from the contracting map [Macleod 1984]. Nevertheless, during initial investigations, and during the development of more accurate algorithms, cubic and quartic formulae can still provide insight and independent verifications [Nievergelt 1994].

Such a sequence of events, from applied problems to abstract mathematics and then further to different applied problems, is typical of the development of mathematics.

Exercises

To simulate the historical experience of encountering a new type of problem, the following six exercises appear here without prior examples or theory.

1. Assuming the usual rules of algebra and that $i^2 = -1$, verify that $(1+2i) \cdot (3+4i) = -5+10i$.

2. Verify that $(1+i) \cdot (1-i) = 2$.

3. Find the dimensions—length and width—of a rectangular altar with a perimeter of 10 yards and an area of 6 square yards.

4. Find the dimensions—length and width—of a hypothetical rectangular altar with a perimeter of 20 yards and an area of 40 square yards.

5. Find the dimensions—length and height—of the rectangular façade of a temple, such that the façade has unit area, and such that the ratio of the height to the length equals twice the ratio of the length to the perimeter of the façade.

6. Consider two parallelipipedic temples, a larger one and a smaller one. Assume that they both have the same shape, in the sense that the dimensions of the larger one are in the same ratios as the dimensions of the smaller one. Also suppose that the larger temple has twice the volume of the smaller temple. Determine the ratio of the height of the larger one to the height of the smaller one.

3. The Algebra of Complex Numbers

Complex numbers admit several rigorous constructions in terms of the set $\mathbb{R}$ of the real numbers, such as the construction that we present here. To understand how one might invent such a construction, examine the origin and the intended use of complex numbers. Historically, complex numbers arose with certain quadratic equations without real solutions, for example, $z^2 - 6z + 13 = 0$, for which the quadratic formula produced an expression of the type $z = 3 \pm \sqrt{-4} = 3 \pm 2\sqrt{-1}$. Such an expression may seem like nonsense, but it may also suggest the construction of new numbers, of the form $u + v\sqrt{-1}$, in terms of two ordinary real numbers, u and v. If two such new numbers lend themselves to the usual rules of arithmetic, for instance, associativity, commutativity, and distributivity, then they should obey the following rules:

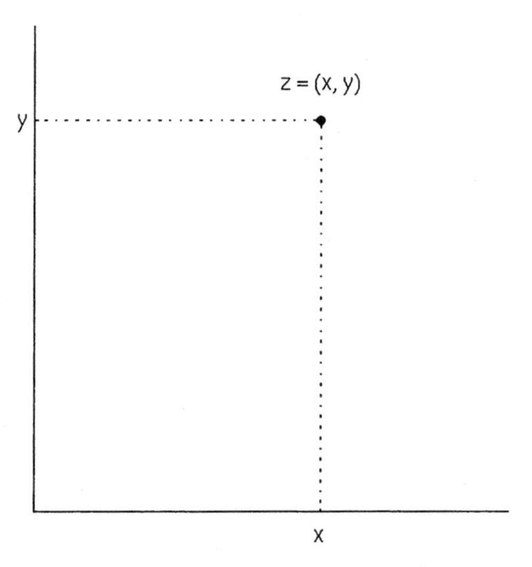

Figure 1. Complex numbers are points or vectors in the Cartesian plane: $\mathbb{C} = \mathbb{R}^2$.

$$\begin{aligned}
(u+v\sqrt{-1})+(x+y\sqrt{-1}) &= (u+x)+(v+y)\sqrt{-1}, \\
(u+v\sqrt{-1})\cdot(x+y\sqrt{-1}) &= (ux+vy\sqrt{-1}\sqrt{-1})+(uy+vx)\sqrt{-1} \\
&= (ux-vy)+(uy+vx)\sqrt{-1}.
\end{aligned}$$

Observe that the specifications just listed involve only arithmetic with the real numbers u, v, x, and y; the symbol $\sqrt{-1}$ serves only to separate u from v, or $(ux-vy)$ from $(uy+vx)$. Such observations suggest the following definition.

Definition 1. A *complex number* is an ordered pair of real numbers, of the form $z=(x,y)$, where $x \in \mathbb{R}$ represents the *first coordinate* and $y \in \mathbb{R}$ represents the *second coordinate* of z. The set of all complex numbers is denoted by $\mathbb{C}$, though it coincides with the plane $\mathbb{R}^2 = \mathbb{R} \times \mathbb{R}$ of all ordered pairs of real numbers. □

With the foregoing definition, the phrases "complex numbers," "vectors," and "points in the real plane" are different names for the same objects: ordered pairs of real numbers [Hahn 1994], as in **Figure 1**. Other—but algebraically equivalent—definitions exist, demonstrated as in subsequent exercises.

Though every complex number consists of a pair of two real numbers, the initial historical lack of a precise definition for complex numbers led to the following terminology.

Definition 2. For each complex number $z=(x,y)$, the first coordinate x is called the *real part* and denoted by $\mathrm{Re}(z)$, whereas the second coordinate is called the *imaginary part* and denoted by $\mathrm{Im}(z)$. For complex numbers expressed by their coordinates, $z=(x,y)$, the usual notation omits one set of parentheses, with $\mathrm{Re}(x,y)$ instead of $\mathrm{Re}((x,y))$ and $\mathrm{Im}(x,y)$ instead of $\mathrm{Im}((x,y))$; thus, $\mathrm{Re}(x,y)=x$ and $\mathrm{Im}(x,y)=y$. □

The set $\mathbb{C}=\mathbb{R}^2$ lends itself to the following two algebraic operations.

Definition 3. The *addition* of complex numbers maps two complex numbers, $w=(u,v)$ and $z=(x,y)$, to their *sum*, $w+z$, defined as the addition of vectors in $\mathbb{R}^2=\mathbb{C}$ by

$$w+z=(u,v)+(x,y):=(u+x,v+y).$$

The *multiplication* of complex numbers maps two complex numbers, $w=(u,v)$ and $z=(x,y)$, to their *product*, denoted by wz, defined by

$$wz=(u,v)(x,y):=(ux-vy,uy+vx).$$ □

Remark 1. In the preceding definition, and in the sequel, expressions of the type $A:=B$ serve to define the yet undefined object on the left (A) by the already defined object on the right (B). Thus the significance of the symbol $:=$ differs from that of $=$, which can designate an identity $A = B$ between two previously defined objects. □

Example 1. Consider the two complex numbers $w := (1, 2)$ and $z := (3, 4)$; then

$$(1,2) + (3,4) = (1+3, 2+4) = (4,6),$$
$$(1,2)(3,4) = ([1 \cdot 3] - [2 \cdot 4],\ [1 \cdot 4] + [2 \cdot 3]) = (3-8,\ 4+6) = (-5, 10).$$

□

Example 2. Consider the two complex numbers $w := (2/3,\ 7/5)$ and $z := (1/7,\ 3/2)$; then

$$(2/3,\ 7/5) + (1/7,\ 3/2) = (2/3 + 1/7,\ 7/5 + 3/2) = (17/21,\ 29/10),$$
$$(2/3,\ 7/5)(1/7,\ 3/2) = ([2/3 \cdot 1/7] - [7/5 \cdot 3/2],\ [2/3 \cdot 3/2] + [7/5 \cdot 1/7])$$
$$= (2/21 - 21/10,\ 1/1 + 1/5) = (-421/210,\ 6/5).$$ □

Example 3. Consider the two complex numbers $w := (\sqrt{2}, \sqrt{3})$ and $z := (\sqrt{5}, \sqrt{7})$; then

$$(\sqrt{2}, \sqrt{3}) + (\sqrt{5}, \sqrt{7}) = (\sqrt{2} + \sqrt{3},\ \sqrt{5} + \sqrt{7}),$$
$$(\sqrt{2}, \sqrt{3})(\sqrt{5}, \sqrt{7}) = ([\sqrt{2} \cdot \sqrt{5}] - [\sqrt{3} \cdot \sqrt{7}],\ [\sqrt{2} \cdot \sqrt{7}] + [\sqrt{3} \cdot \sqrt{5}]).$$ □

Example 4. Consider the two complex numbers $w := (\pi, \sqrt{2})$ and $z := (5/7, e)$; then

$$(\pi, \sqrt{2}) + (5/7, e) = (\pi + 5/7,\ \sqrt{2} + e),$$
$$(\pi, \sqrt{2})(5/7, e) = ([\pi \cdot 5/7] - [\sqrt{2} \cdot e],\ [\pi \cdot e] + [\sqrt{2} \cdot 5/7]).$$ □

Geometrically, the complex addition coincides with the addition of vectors in the plane: for every pair of complex numbers w and z, their sum $w+z$ lies at the fourth vertex of the parallelogram with vertices at $(0,0)$, w, z, and $w+z$. In contrast, the complex multiplication corresponds to a rotation followed by a dilation or a compression, as explained in the next section. Algebraically, however, both operations satisfy the same usual rules of algebra as do the rational and the real numbers.

Theorem 1. *The addition and the multiplication of complex numbers obey the algebraic rules listed in* **Table 1**.

Proof: For each equality, calculate the left-hand side and the right-hand side separately with the definition of the addition and of the multiplication, and verify that the results agree with each other. For example, the proof of the associativity of the multiplication proceeds as follows.

Table 1.

Algebraic properties of the complex numbers.

The following properties hold for all complex numbers (u, v), (x, y), and (p, q).

(1) Associativity of $+$	$[(u,v)+(x,y)]+(p,q)=(u,v)+[(x,y)+(p,q)]$
(2) Commutativity of $+$	$(u,v)+(x,y)=(x,y)+(u,v)$
(3) Additive identity	$(x,y)+(0,0)=(x,y)=(0,0)+(x,y)$
(4) Additive inverse	$(x,y)+(-x,-y)=(0,0)$
(5) Associativity of $\cdot$	$[(u,v)(x,y)](p,q)=(u,v)[(x,y)(p,q)]$
(6) Commutativity of $\cdot$	$(u,v)(x,y)=(x,y)(u,v)$
(7) Multiplicative identity	$(x,y)(1,0)=(x,y)=(1,0)(x,y)$
(8) Multiplicative inverse	If $(x,y)\neq 0$, then $(x,y)(x/[x^2+y^2],-y/[x^2+y^2])=(1,0)$
(9) Distributivity	$(u,v)[(x,y)+(p,q)]=[(u,v)(x,y)]+[(u,v)(p,q)]$

$$\begin{aligned}
[(u,v)(x,y)](p,q) &= (ux-vy, uy+vx)(p,q)\\
&= ([ux-vy]p-[uy+vx]q,\ [ux-vy]q+[uy+vx]p)\\
&= (uxp-vyp-uyq-vxq,\ uxq-vyq+uyp+vxp);
\end{aligned}$$

$$\begin{aligned}
(u,v)[(x,y)(p,q]) &= (u,v)(xp-yq, xq+yp)\\
&= (u[xp-yq]-v[xq+yp],\ u[xq+yp]+v[xp-yq])\\
&= (uxp-uyq-vxq-vyp,\ uxq+uyp+vxp-vyq)\\
&= (uxp-vyp-uyq-vxq,\ uxq-vyq+uyp+vxp)\\
&= [(u,v)(x,y)](p,q). \quad \checkmark
\end{aligned}$$

The proofs of the other properties proceed similarly and form the object of exercises. □

Remark 2. *(Inclusion of* $\mathbb{R}$ *into* $\mathbb{C}$*.)* The set of all complex numbers with second coordinate equal to zero, of the form $(x, 0)$, satisfies all the algebraic rules of the real numbers with ordinary addition and multiplication:

$$\begin{aligned}
(x,0)+(u,0) &= (x+u,0),\\
(x,0)(u,0) &= (xu,0).
\end{aligned}$$

So, we identify the set $\mathbb{R}$ of all real numbers with the set $\{(x,0) : x \in \mathbb{R}\} \subset \mathbb{C}$. With such an identification, $1(p,q)$ means $(1,0)(p,q)$, and, similarly, $0+(p,q)$ means $(0,0)+(p,q)$. The additive identity, $(0,0)$, is also called the *origin.* More generally, for each real number r, $r(p,q)$ means $(r,0)(p,q)$, and, similarly, $r+(p,q)$ means $(r,0)+(p,q)$. Thus,

$$\begin{aligned}
r(p,q) &= (r,0)(p,q) = (rp-0q, rq+0p) = (rp, rq),\\
r+(p,q) &= (r,0)+(p,q) = (r+p,q).
\end{aligned}$$

□

Example 5. The complex number $(0,1)$ has a negative square:

$$(0,1)(0,1) = (0 \cdot 0 - 1 \cdot 1,\ 0 \cdot 1 + 1 \cdot 0) = (-1,0) = -1. \qquad \square$$

Thus, $(0,1)^2 = -1$, and, consequently, we denote $(0,1)$ by the symbols $\sqrt{-1}$ or i, as the first letter of the word "imaginary" [Hahn 1994]. A subsequent section will show that multiplication by -1 corresponds to a rotation by one half of a turn, and that $(0,1)$ corresponds to a rotation by one quarter of a turn, with complex multiplication corresponding to the composition of transformations.

In **Table 1**, the formula for the multiplicative inverse,

$$(x,y)^{-1} = \left(\frac{x}{x^2+y^2}, \frac{-y}{x^2+y^2}\right) = \frac{1}{x^2+y^2}\,(x,-y),$$

also helps in dividing a complex number by another.

Definition 4.

$$\frac{(u,v)}{(x,y)} := (u,v)\,(x,y)^{-1} = (x,y)^{-1}(u,v).$$

Example 6. To divide $(1,2)$ by $(3,4)$, first determine the multiplicative inverse $(3,4)^{-1}$:

$$(3,4)^{-1} = \left(\frac{3}{3^2+4^2}, \frac{-4}{3^2+4^2}\right) = \frac{1}{3^2+4^2}\,(3,-4) = \frac{1}{25}\,(3,-4).$$

As a verification of the result just obtained, check that $(3,4)^{-1}(3,4) = 1$:

$$\begin{aligned}\left(\frac{3}{3^2+4^2}, \frac{-4}{3^2+4^2}\right)(3,4) &= \frac{1}{25}\,(3,-4)\,(3,4)\\ &= \frac{1}{25}\,(3\cdot 3-(-4)\cdot 4, 3\cdot 4+(-4)\cdot 3)\\ &= \frac{1}{25}\,(25,0) = (1,0) = 1. \quad \checkmark\end{aligned}$$

Finally, to divide $(1,2)$ by $(3,4)$, use $(3,4)^{-1}$ as follows:

$$\begin{aligned}\frac{(1,2)}{(3,4)} &= (3,4)^{-1}(1,2) = \frac{1}{25}\,(3,-4)\,(1,2)\\ &= \frac{1}{25}\,(3\cdot 1-(-4)\cdot 2,\ 3\cdot 2+(-4)\cdot 1) = \frac{1}{25}\,(11,2) = \left({}^{11}\!/_{25},\ {}^{2}\!/_{25}\right).\end{aligned}$$

As a verification, check that $(3,4)\left({}^{11}\!/_{25},\ {}^{2}\!/_{25}\right) = (1,2)$. $\square$

Remark 3. *(Relations with abstract algebra.)* The complex numbers with the complex addition and multiplication share many algebraic properties with other systems of numbers. The first four properties in **Table 1** mean that $(\mathbb{C}, +)$ is a *commutative group.* Another example of a commutative group consists of the set $\mathbb{Z} := \{\dots, -2, -1, 0, 1, 2, \dots\}$ of all the integers, positive and negative, with the ordinary addition $+$. Similarly, properties (5)–(8) mean that $(\mathbb{C}, \cdot)$ is a *commutative monoid.* The set $\mathbb{N} := \{0, 1, 2, 3, \dots\}$ of all nonnegative integers gives rise to two commutative monoids: $(\mathbb{N}, \cdot)$ with multiplication, and $(\mathbb{N}, +)$ with addition. Finally, properties (1)–(9) mean that $(\mathbb{C}, +, \cdot)$ is a *field of numbers* ("field," for short). Other examples of fields include, for instance, the rational numbers $(\mathbb{Q}, +, \cdot)$ and the real numbers $(\mathbb{R}, +, \cdot)$ [Birkhoff and Mac Lane 1977; Hungerford 1974; Lang 1965]. In contrast, $(\mathbb{Z}, +, \cdot)$ is not a field, because it does not contain the multiplicative inverse of all integers; for instance, $1/2 \notin \mathbb{Z}$. □

Exercises

7. Evaluate the following expressions.

a) $(3,2)+(1,4)$
b) $(1,4)+(3,2)$
c) $(3,2)(1,4)$
d) $(1,4)(3,2)$
e) $(4,3)[(3,2)+(1,4)]$
f) $[(4,3)(3,2)]+[(4,3)(1,4)]$
g) $(4,3)^{-1}$
h) $[(3,2)+(1,4)]+(9,5)$
i) $(3,2)+[(1,4)+(9,5)]$
j) $[(3,2)(1,4)](9,5)$
k) $(3,2)[(1,4)(9,5)]$
l) $-(4,3)=$
m) $(3,2)-(4,3)=$
n) $(3,2)/(4,3)=$

8. Evaluate the following expressions.

a) $(7,1)+(2,5)$
b) $(2,5)+(7,1)$
c) $(7,1)(2,5)$
d) $(2,5)(7,1)$
e) $(5,12)[(7,1)+(2,5)]$
f) $[(5,12)(7,1)]+[(5,12)(2,5)]$
g) $(5,12)^{-1}$
h) $[(7,1)+(2,5)]+(3,3)$
i) $(7,1)+[(2,5)+(3,3)]$
j) $[(7,1)(2,5)](3,3)$
k) $(7,1)[(2,5)(3,3)]$
l) $-(5,12)$
m) $(7,1)-(5,12)$
n) $(7,1)/(5,12)$

9. Complete the proof of **Theorem 1**, that the addition and multiplication of complex numbers satisfy each property in **Table 1**.

10. Verify that $(0,0)(a,b)=(0,0)$ for every complex number (a,b).

11. Verify that $(-x,-y)+(x,y)=(0,0)$, so that the same additive inverse works on the left and on the right.

12. Verify that $(x/[x^2+y^2], -y/[x^2+y^2]) \cdot (x, y) = (1, 0)$, so that the same multiplicative inverse works on the left and on the right.

13. Explain how one may discover the formula for the multiplicative inverse $(x, y)^{-1}$ by solving for u and v the real and complex parts of the equation $(u, v)(x, y) = (1, 0)$.

14. A *geometric series* is a sum of powers, of the type $1+z+z^2+z^3+\cdots+z^{n-1}+z^n$. For each positive integer n, and for each complex number $z \neq 1$, prove that

$$1+z+z^2+z^3+\cdots+z^{n-1}+z^n = \frac{1-z^{n+1}}{1-z}.$$

15. (†) Determine whether a formula exists that expresses the complex multiplication of two complex numbers with only *three real* multiplications (but any number of additions and subtractions) of their coordinates. In contrast, the definition $(u, v)(x, y) = (ux - vy, uy + vx)$ involves one addition, one subtraction, and *four* real multiplications: ux, vy, uy, and vx.

16. Denote by $\mathbb{Q}(i)$ the set of all pairs of rational numbers:

$$\mathbb{Q}(i) := \{(r, s) : r, s \in \mathbb{Q}\}.$$

For all rational numbers $r, s, p, q \in \mathbb{Q}$, define an addition and a multiplication on the set $\mathbb{Q}(i)$ by

$$(r, s) + (p, q) := (r + p, s + q),$$
$$(r, s)(p, q) := (rp - sq, rq + sp).$$

Determine which of the algebraic properties in **Table 2** hold, if any, and which do not hold, if any.

Table 2.

Which of the following properties hold for all pairs (u, v), (p, q), and (b, d)?

(1) Associativity of +	$[(u,v)+(p,q)]+(b,d) = (u,v)+[(p,q)+(b,d)]$
(2) Commutativity of +	$(u,v)+(p,q) = (p,q)+(u,v)$
(3) Additive identity	$(p,q)+(0,0) = (p,q) = (0,0)+(p,q)$
(4) Additive inverse	$(p,q)+(-p,-q) = (0,0)$
(5) Associativity of $\cdot$	$[(u,v)(p,q)](b,d) = (u,v)[(p,q)(b,d)]$
(6) Commutativity of $\cdot$	$(u,v)(p,q) = (p,q)(u,v)$
(7) Multiplicative identity	$(p,q)(1,0) = (p,q) = (1,0)(p,q)$
(8) Multiplicative inverse	If $(p,q) \neq 0$, then $(p,q)(p/[p^2+q^2], -q/[p^2+q^2]) = (1,0)$
(9) Distributivity	$(u,v)[(p,q)+(b,d)] = [(u,v)(p,q)]+[(u,v)(b,d)]$

17. Denote by $\mathbb{Z}(i)$ the set of all pairs of integers:

$$\mathbb{Z}(i) := \{(r, s) \, : \, r, s \in \mathbb{Z}\}.$$

For all integers $r, s, p, q \in \mathbb{Z}$, define an addition and a multiplication on the set $\mathbb{Z}(i)$ by

$$\begin{aligned}(r, s) + (p, q) &:= (r + p, s + q),\\ (r, s)(p, q) &:= (rp - sq, rq + sp).\end{aligned}$$

Determine which of the algebraic properties in **Table 2** hold, if any, and which do not hold.

18. This exercise outlines another construction of the complex numbers, not through the real plane, but through a special set of real matrices. To this end, let $\mathbb{M}_{2\times 2}(\mathbb{R})$ denote the set of all matrices with real entries in two rows and two columns, with addition and multiplication of matrices defined as in linear algebra [Jacob 1990]:

$$\begin{pmatrix} p & q \\ g & y \end{pmatrix} + \begin{pmatrix} h & k \\ \ell & t \end{pmatrix} = \begin{pmatrix} p+h & q+k \\ g+\ell & y+t \end{pmatrix},$$

$$\begin{pmatrix} p & q \\ g & y \end{pmatrix} \begin{pmatrix} h & k \\ \ell & t \end{pmatrix} = \begin{pmatrix} ph+q\ell & pk+qt \\ gh+y\ell & gk+yt \end{pmatrix}.$$

Consider the subset $C \subset \mathbb{M}_{2\times 2}(\mathbb{R})$ of all matrices of the form

$$\begin{pmatrix} a & b \\ -b & a \end{pmatrix},$$

and consider the function $M : \mathbb{C} \to C$, which maps each complex number (a, b) to the matrix $M(a, b)$ defined by

$$M(a, b) := \begin{pmatrix} a & b \\ -b & a \end{pmatrix}.$$

Prove that the function M satisfies the following properties.

a) M is injective: if $M(a, b) = M(c, d)$, then $(a, b) = (c, d)$.

b) M is surjective: for each $A \in C$, some $z \in \mathbb{C}$ exists for which $A = M(z)$.

c) M is a "homomorphism of additive groups": $M((a, b) + (c, d)) = M(a, b) + M(c, d)$ for each (a, b) and each (c, d) in $\mathbb{C}$.

d) M is a "homomorphism of multiplicative monoids": $M((a, b)(c, d)) = M(a, b)M(c, d)$ for each (a, b) and each (c, d) in $\mathbb{C}$.

e) M is an "isomorphism of fields": $M(0, 0)$ has all entries equal to zero, $M(-a, -b) = -M(a, b)$, $M\left((a, b)^{-1}\right) = (M(a, b))^{-1}$, and $M(1, 0)$ is the identity matrix I, defined by

$$I = \begin{pmatrix} 1 & 0 \\ 0 & 1 \end{pmatrix}.$$

The foregoing property, while verifiable independently, as suggested here, also follows logically from the preceding properties [Hungerford 1974, #1 p. 33 and #15 p. 121].

f) All of the preceding properties also hold for the restriction of M to the set of "real numbers" $\{(x, 0) : x \in \mathbb{R}\} \subset \mathbb{C}$ into the subset of all diagonal matrices in C.

Thus, C satisfies all the algebraic properties of $\mathbb{C}$ and could also serve as complex numbers.

19. The set $\mathbb{H} := \mathbb{R}^4$ of quaternions consists of all quadruples (x, y, z, t) of real numbers, with addition defined by coordinates,

$$(x_1, y_1, z_1, t_1) + (x_2, y_2, z_2, t_2) := (x_1 + x_2, y_1 + y_2, z_1 + z_2, t_1 + t_2),$$

and an associative multiplication that distributes on the left of addition, defined by the specification

$$i^2 = j^2 = k^2 = ijk = -1$$

with the multiplicative identity $1 := (1, 0, 0, 0)$ and with

$$i := (0, 1, 0, 0), \qquad j := (0, 0, 1, 0), \qquad k := (0, 0, 0, 1).$$

Thus, the product $p = gq$ of two quaternions g and q takes the form

$$\begin{aligned}
g &= (g_1, g_2, g_3, g_4), \\
q &= (q_1, q_2, q_3, q_4), \\
p &= (p_1, p_2, p_3, p_4), \\
p_1 &= g_1 q_1 - g_2 q_2 - g_3 q_3 - g_4 q_4, \\
p_2 &= g_1 q_2 - g_2 q_1 - g_3 q_4 - g_4 q_3, \\
p_3 &= g_1 q_3 - g_2 q_4 - g_3 q_1 - g_4 q_2, \\
p_4 &= g_1 q_4 - g_2 q_3 - g_3 q_2 - g_4 q_1.
\end{aligned}$$

Prove that the quaternions form a real noncommutative division-algebra (a field where the multiplication does not commute, containing the real numbers in its center).

20. (†) Kustaanheimo and Stiefel's method (called the *KS method*) [1971] to regularize the three-dimensional differential equations of orbital mechanics involves the following algebra on the four-dimensional Euclidean space $\mathbb{R}^4$. For each vector $\vec{\mathbf{u}} \in \mathbb{R}^4$, define the matrix

$$L(\vec{\mathbf{u}}) := \begin{pmatrix} u_1 & -u_2 & -u_3 & u_4 \\ u_2 & u_1 & -u_4 & -u_3 \\ u_3 & u_4 & u_1 & u_2 \\ u_4 & -u_3 & u_2 & -u_1 \end{pmatrix};$$

then define the product of two vectors $\vec{u}$ and $\vec{v}$ by the product of matrices

$$\vec{u} \star \vec{v} := L(\vec{u})\vec{v}.$$

Investigate the associativity and the commutativity of the product $\star$. Also investigate whether $\star$ distributes on the left and on the right of the ordinary addition of vectors in $\mathbb{R}^4$. Moreover, investigate whether $(\mathbb{R}^4, +, \star)$ is an algebra over $\mathbb{R}$, or a division-ring, or a field.

Finally, consider a subset S of Levi-Civita's type in $\mathbb{R}^4$, defined by the property that for each $\vec{u}$ and each $\vec{v}$ in S the following relation holds:

$$u_4 v_1 - u_3 v_2 + u_2 v_3 - u_1 v_4 = 0.$$

Investigate whether the restriction of $\star$ to such a set S is associative or commutative. Determine whether such a set S forms a field.

4. The Geometry of Complex Numbers

The definition of a complex number as an ordered pair $z = (x, y)$ of real numbers yields a geometric representation of the complex number z as a point (x, y) in the plane $\mathbb{C} = \mathbb{R}^2$. Such a geometric representation may provide a visual intuition and a better understanding for problems involving complex numbers.

4.1 Lengths

One of the most elementary concepts relating algebra to geometry consists of the ordinary length, also called the magnitude or the "modulus," of a complex number, and defined as the distance from the origin to that complex number.

Definition 5. The *modulus*—also called the *length*, or the *magnitude*, or the *Euclidean norm*—of a complex number $z = (x, y)$ is the *real* number $|z|$ defined by the nonnegative square root

$$|(x, y)| := \sqrt{x^2 + y^2}.$$

Example 7. $|(3, 4)| = \sqrt{3^2 + 4^2} = \sqrt{9 + 16} = \sqrt{25} = 5$. □

Example 8. $|(20, 21)| = \sqrt{20^2 + 21^2} = \sqrt{400 + 441} = \sqrt{841} = 29$. □

The modulus gives a first geometric interpretation of the complex multiplication: the modulus of the product equals the product of the moduli.

Proposition 1. *For all complex numbers* $w = (u, v)$ *and* $z = (x, y)$,
$$|w\,z| = |w|\,|z|.$$

Proof: Calcumate each side with the definitions, and compare the results to each other:

$$\begin{aligned}
|w\,z|^2 &= |(u,v)(x,y)|^2 = |(ux - vy,\ uy + vx)|^2 \\
&= (ux - vy)^2 + (uy + vx)^2 \\
&= u^2x^2 - 2uxvy + v^2y^2 + u^2y^2 + 2uyvx + v^2x^2 \\
&= u^2x^2 + v^2y^2 + u^2y^2 + v^2x^2, \\
(|w|\,|z|)^2 &= |w|^2|z|^2 = (u^2 + v^2)(x^2 + y^2) = u^2x^2 + u^2y^2 + v^2x^2 + v^2y^2 \\
&= |w\,z|^2. \quad \checkmark
\end{aligned}$$

Hence, taking real square roots gives $|w\,z| = |w|\,|z|$. □

Example 9. Let $w := (3,4)$ and $z := (20,21)$. The preceding examples have shown that $|w| = 5$ and $|z| = 29$. Moreover,

$$\begin{aligned}
w\,z &= (3,4)(20,21) = ([3 \cdot 20] - [4 \cdot 21],\ [3 \cdot 21] + [4 \cdot 20]) \\
&= (60 - 84,\ 63 + 80) = (-24,\ 143), \\
|w\,z| &= |(-24,\ 143)| = \sqrt{(-24)^2 + 143^2} = \sqrt{567 + 20{,}449} = \sqrt{21{,}025} \\
&= 145 = 5 \cdot 29 = |w|\,|z|. \quad \checkmark
\end{aligned}$$

As in Euclidean geometry, the modulus of complex numbers satisfies the "triangle inequality," which states that the length of each side of a triangle does not exceed the sum of the lengths of the other two sides.

Theorem 2. (Triangle Inequality.) *For all complex numbers* $w = (u, v)$ *and* $z = (x, y)$,
$$|w + z| \leq |w| + |z|.$$
Moreover, the equality $|w + z| = |w| + |z|$ *holds if, but only if, a real number* r *exists for which* $w = rz$ *or* $z = rw$, *which means that* w *and* z *lie on the same half-line from the origin.*

Proof: To discover a proof, square both sides of the proposed inequality
$$|w + z|^2 \leq (|w| + |z|)^2$$
and apply the definition of the modulus to both sides:

$$\begin{aligned}
|w + z| &\leq |w| + |z|, \\
|w + z|^2 &\leq (|w| + |z|)^2, \\
|(u + x, v + y)|^2 &\leq |(u,v)|^2 + 2|(u,v)|\,|(x,y)| + |(x,y)|^2, \\
(u^2 + 2ux + x^2) + (v^2 + 2vy + y^2) &\leq u^2 + v^2 + 2\sqrt{u^2 + v^2}\sqrt{x^2 + y^2} + x^2 + y^2, \\
ux + vy &\leq \sqrt{u^2 + v^2}\sqrt{x^2 + y^2},
\end{aligned}$$

$$
\begin{aligned}
u^2x^2 + 2uxvy + v^2y^2 &\leq u^2x^2 + u^2y^2 + v^2x^2 + v^2y^2, \\
0 &\leq u^2y^2 - 2uxvy + v^2x^2, \\
0 &\leq (uy - vx)^2.
\end{aligned}
$$

Because the real square $(uy - vx)^2$ cannot be negative, the last inequality holds. Hence, because all the inequalities just obtained are equivalent to one another, then the proof just discovered proceeds in the reverse direction, from the last inequality, $0 \leq (uy - vx)^2$, to the first inequality, $|w + z| \leq |w| + |z|$, which then also holds. Moreover, by equivalence of the preceding inequalities, the equality $|w + z| = |w| + |z|$ holds if, but only if, $0 = (uy - vx)^2$ which holds if, but only if, $uy = vx$. If $x \neq 0$ and $y \neq 0$, $uy = vx$ means that $u/x = v/y$; setting $r := u/x = v/y$ then shows that $u = rx$ and $v = ry$, whence $w = (u, v) = r(x, y) = rz$. Similar reasonings again give $w = rz$ or $z = rw$ in the other cases. □

Example 10. Let $w := (3, 0)$ and $z := (0, 4)$. Then $|w| = \sqrt{3^2 + 0^2} = 3$ and $|z| = \sqrt{0^2 + 4^2} = 4$. However, $w + z = (3, 0) + (0, 4) = (3, 4)$ and $|w + z| = |(3, 4)| = 5$. Thus,

$$|(3, 0) + (0, 4)| = 5 < 3 + 4 = |w| + |z|. \quad \checkmark$$

Also as in Euclidean geometry, the modulus of complex numbers satisfies the "reverse triangle inequality," which states that the difference of the lengths of any two sides of a triangle does not exceed the length of the third side.

Theorem 3. **(Reverse Triangle Inequality.)** *For all complex numbers* $w = (u, v)$ *and* $z = (x, y)$,

$$|w - z| \geq \big|\, |w| - |z| \,\big|.$$

Proof: The proof forms the object of an exercise. □

Example 11. Let $w := (3, 0)$ and $z := (0, 4)$. Then $|w| = \sqrt{3^2 + 0^2} = 3$ and $|z| = \sqrt{0^2 + (-4)^2} = 4$, whence $\big|\, |w| - |z| \,\big| = |3 - 4| = |-1| = 1$. However, $w - z = (3, 0) - (0, 4) = (3, -4)$ and $|w + z| = |(3, -4)| = 5$. Thus, as with a right triangle with sides of lengths $3, 4, 5$,

$$|(3, 0) - (0, 4)| = 5 > 1 = \big|\, |(3, 0)| - |(0, 4)| \,\big|. \quad \checkmark$$

Another elementary concept relating algebra to geometry consists of the "complex conjugate" of a complex number, which corresponds to a symmetry across the first coordinate axis.

Definition 6. The *complex conjugate* of a complex number $z = (x, y)$ is the complex number

$$\overline{z} := (x, -y). \qquad \square$$

Example 12. $\overline{(3, 4)} = (3, -4)$. □

Proposition 2. *The complex conjugate satisfies the properties in* **Table 3**.

Table 3.
Properties of the complex conjugate.

(C1) Additivity	$\overline{w+z}$	$=$	$\overline{w}+\overline{z}$
(C2) Multiplicativity	$\overline{w\,z}$	$=$	$\overline{w}\,\overline{z}$
(C3)	$z+\overline{z}$	$=$	$2\mathrm{Re}(z)$
(C4)	$z-\overline{z}$	$=$	$2i\mathrm{Im}(z)$
(C5)	$z\,\overline{z}$	$=$	$\lvert z\rvert^2$
(C6)	$\overline{1/z}$	$=$	$1/\overline{z}$

Proof: The proof consists of routine verifications, left to the exercises. □

Exercises

21. Evaluate the following moduli.

a) $|(12,0)|$
b) $|(0,5)|$
c) $|(12,0)+(0,5)|$
d) $|(12,0)|+|(0,5)|$
e) $|(12,0)(0,5)|$
f) $|(12,0)-(0,5)|$
g) $|\,|(12,0)|-|(0,5)|\,|$
h) $|(4,3)|$
i) $|(12,5)|$
j) $|(12,5)(4,3)|$

22. Evaluate the following moduli.

a) $|(8,0)|$
b) $|(0,15)|$
c) $|(8,0)+(0,15)|$
d) $|(8,0)|+|(0,15)|$
e) $|(8,0)(0,15)|$
f) $|(8,0)-(0,15)|$
g) $|\,|(8,0)|-|(0,15)|\,|$
h) $|(8,15)|$
i) $|(7,24)|$
j) $|(8,15)(7,24)|$

23. Prove the reverse triangle inequality. The proof need not resemble that of the triangle inequality.

24. Prove that $z=\overline{z}$ if, but only if, $z\in\mathbb{R}$.

25. Prove the properties of the complex conjugate in **Table 3**.

26. Prove that if $z\neq 0$, then $|z/\overline{z}|=1$.

27. Prove that if two integers p and q are each the sum of two squared integers, so that $p=m^2+n^2$ and $q=k^2+\ell^2$, then so is their product pq.

28. Investigate whether if two integers p and q are each the sum of two squared integers, so that $p = m^2 + n^2$ and $q = k^2 + \ell^2$, then so is their sum $p + q$.

29. This exercise shows that the complex roots of a real polynomial occur only in conjugate pairs. Consider *real* numbers $a_0, \dots, a_n$, and suppose that z is a *complex* number such that $a_0 + a_1 z + \cdots + c_{n-1} z^{n-1} + c_n z^n = 0$. Prove that $a_0 + a_1 \overline{z} + \cdots + c_{n-1} \overline{z}^{n-1} + c_n \overline{z}^n = 0$.

30. Prove that if the numbers $c_1, c_2, \dots, c_{n-1}, c_n$ are all nonnegative, and if c_0 and at least one c_k for $k > 0$ are both positive, then the equation $c_n v^n + c_{n-1} v^{n-1} + c_{n-2} v^{n-2} + \cdots + c_2 v^2 + c_1 v - c_0 = 0$ has exactly one positive solution.

The following two exercises provide estimates—with only real arithmetic or real roots—of the locations of the roots of polynomials.

31. For each polynomial p defined by $p(z) = c_n z^n + \cdots + c_2 z^2 + c_1 z + c_0$, with complex coefficients $c_n, \dots, c_0$, prove that all the roots of p lie in the region where

$$\frac{|c_0|}{\min\{|c_\ell| + |c_0| \,:\, 0 < \ell \leq n\}} \leq |z| \leq 1 + \frac{\max\{|c_k| \,:\, 0 \leq k < n\}}{|c_n|}.$$

32. For each polynomial p defined by $p(z) = c_n z^n + \cdots + c_2 z^2 + c_1 z + c_0$, with complex coefficients $c_n, \dots, c_0$, prove that all the roots of p lie in the region where

$$\min\left\{\left(\frac{|c_0|}{n|c_\ell|}\right)^{1/\ell} \,:\, 1 \leq \ell \leq n\right\} \leq |z| \leq \max\left\{\left(\frac{n|c_{n-\ell}|}{|c_n|}\right)^{1/\ell} \,:\, 1 \leq \ell \leq n\right\}.$$

4.2 Angles

While the modulus of a complex number measures the distance from that complex number to the origin, another quantity, called the "argument," measures the "angle" from the first coordinate axis to that complex number, as in **Figure 2**.

Remark 4. Because an angle measures the "length" of an arc of a circle with unit radius between two straight rays, a rigorous definition of the concept of "angle" requires such an infinite iterative process as a Riemann integral or an infinite series, as in calculus, or separate axioms about the measure of angles [Kelly and Mathews 1981, 31]. Consequently, either we may assume such prerequisites from previous courses, or we may consider the present subsection about angles as an informal aid for visual intuition and understanding. □

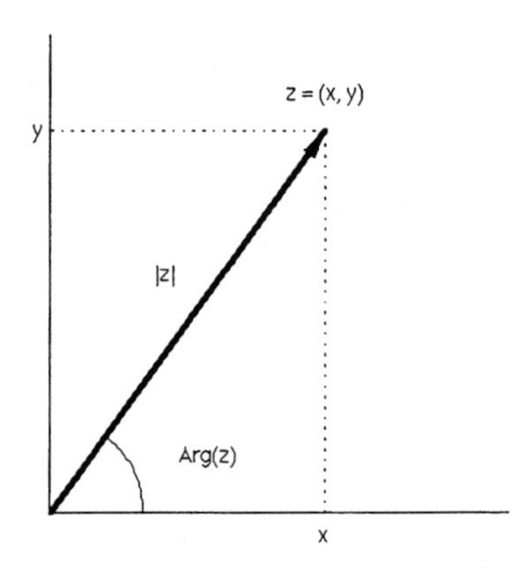

Figure 2. Polar coordinates of complex numbers are polar coordinates in the plane.

Definition 7. The *principal argument* of a nonzero complex number z is the real number, denoted by $\operatorname{Arg}(z)$, which measures the angle from the positive real axis to z, counterclockwise, and so that $-\pi < \operatorname{Arg}(z) \leq \pi$. For complex numbers expressed by their coordinates, $z = (x, y)$, the usual notation omits one set of parentheses, with $\operatorname{Arg}(x, y)$ instead of $\operatorname{Arg}((x, y))$. □

Example 13. For every *positive real* number $x > 0$, $\operatorname{Arg}(x, 0) = 0$, because $(x, 0)$ lies on the positive real axis. □

Example 14. For every *negative real* number $t < 0$, $\operatorname{Arg}(t, 0) = \pi$, because $(t, 0)$ lies one half of a turn counterclockwise from the real axis. (The point $(t, 0)$ also lies one half of a turn clockwise from the real axis, but $-\pi$ does not lie in the range $]-\pi, \pi]$ of the principal argument, which forces the choice of π for $\operatorname{Arg}(t, 0)$.) □

Example 15. For every positive real number $r > 0$,

$$\operatorname{Arg}(ri) = \operatorname{Arg}(0, r) = \pi/2,$$

because $ri = (0, r)$ lies one quarter of a turn counterclockwise from the positive real axis. □

Example 16. For every negative real number $s < 0$,

$$\operatorname{Arg}(si) = \operatorname{Arg}(0, s) = -\pi/2,$$

because $si = (0, s)$ lies one quarter of a turn clockwise from the positive real axis. □

Remark 5. Other "arguments" exist, because adding any integral multiple of 2π to an angle merely adds an integral number of complete rotations and thus produces the same complex number. For example, another common choice of the argument lies in the range $[0, 2\pi[$ from 0 included to 2π excluded. (In the present context, open square brackets avoid the confusion of an open interval $]a, b[$ with a complex numbers (a, b)). The principal argument serves as a convenient convention to ensure agreement among users, which proves crucial in the solution of certain practical problems [Hewlett-Packard 1984, 70; Kahan 1987]. □

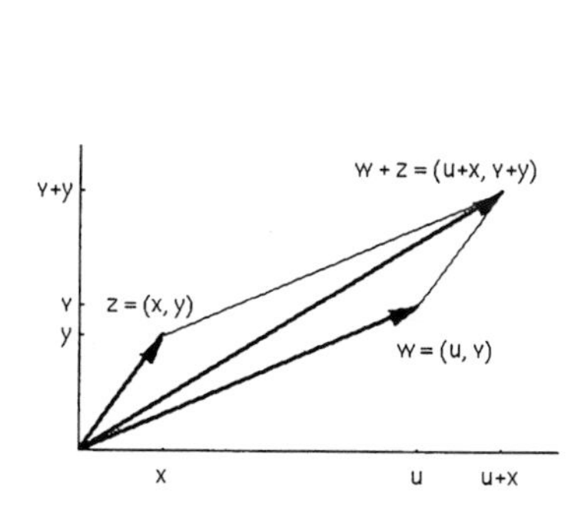

Figure 3. The sum of two complex numbers.

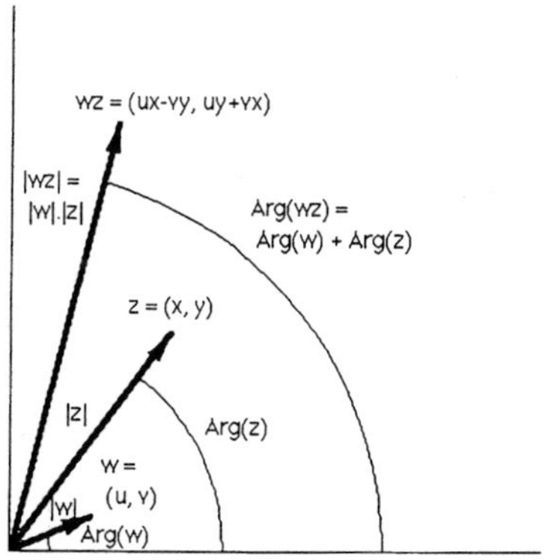

Figure 4. The product of two complex numbers.

While vectors provide a geometric interpretation of the complex addition, as in **Figure 3**, the "argument" provides a geometric interpretation of the complex multiplication: the argument of the product equals the *sum* of the arguments, as in **Figure 4**, plus an integral multiple of 2π.

Proposition 3. *For all complex numbers* $w = (u, v)$ *and* $z = (x, y)$, *an integer* k *exists for which* $\operatorname{Arg}(w\,z) = \operatorname{Arg}(w) + \operatorname{Arg}(z) + 2k\pi$.

Proof: Use the trigonometric identity for the tangent of a sum:

$$\begin{aligned}\tan(\operatorname{Arg}(w) + \operatorname{Arg}(z)) &= \frac{\tan(\operatorname{Arg}(w) + \operatorname{Arg}(z))}{1 - \tan(\operatorname{Arg}(w))\tan(\operatorname{Arg}(z))} \\ &= \frac{(v/u) + (y/x)}{1 - (v/u)(y/x)} = \frac{vx + uy}{ux - vy} \\ &= \frac{\operatorname{Im}(wz)}{\operatorname{Re}(wz)} = \tan(\operatorname{Arg}(wz)).\end{aligned}$$

Thus, $\tan(\operatorname{Arg}(w) + \operatorname{Arg}(z)) = \tan(\operatorname{Arg}(wz))$. Similar calculations apply to the sine and cosine, whence $\operatorname{Arg}(w\,z) = \operatorname{Arg}(w) + \operatorname{Arg}(z) + 2k\pi$. □

Whereas the real and imaginary parts, x and y, are called the *Cartesian coordinates,* the modulus and the argument, $|z|$ and $\operatorname{Arg}(z)$, are called the *polar coordinates,* also often denoted by $r = |z|$ and $\theta = \operatorname{Arg}(z)$, where θ represents the Greek letter "theta." The right triangle with vertices at $(0,0)$, $(x,0)$, and (x,y) lead to the following formulae to change coordinates:

$$\begin{cases} x &= \operatorname{Re}(z) = |z|\cos(\operatorname{Arg}(z)) = r\cos\theta, \\ y &= \operatorname{Im}(z) = |z|\sin(\operatorname{Arg}(z)) = r\sin\theta, \end{cases}$$

$$\begin{cases} r = |z| &= \sqrt{x^2+y^2}, \\ \cos\theta = \cos(\operatorname{Arg}(z)) &= x/\sqrt{x^2+y^2}, \\ \sin\theta = \sin(\operatorname{Arg}(z)) &= y/\sqrt{x^2+y^2}. \end{cases}$$

Example 17. If $z = (\sqrt{3},1)$, then $|z| = \sqrt{3+1} = 2$, $\cos(\operatorname{Arg}(z)) = \sqrt{3}/2$, $\sin(\operatorname{Arg}(z)) = 1/2$, whence $\theta = \operatorname{Arg}(z) = \pi/6$. □

Example 18. If $r = |w| = 5$ and $\theta = \operatorname{Arg}(w) = -\pi/12$, then

$$\cos(\operatorname{Arg}(z)) = \cos(-\pi/12) = \sqrt{2+\sqrt{3}}\,/2$$

and

$$\sin(\operatorname{Arg}(z)) = \sin(-\pi/12) = -\sqrt{2-\sqrt{3}}\,/2.$$

Hence, $x = \operatorname{Re}(w) = 5\sqrt{2+\sqrt{3}}\,/2$ and $y = \operatorname{Im}(w) = -5\sqrt{2-\sqrt{3}}\,/2$. □

The additive feature of the argument function provides a convenient means to express rotations in the plane. Indeed, if a complex number w has unit modulus and argument θ, so that $|w| = 1$ and $\operatorname{Arg}(w) = \theta$, whence $w = \cos\theta + i\sin\theta$, then for each $z \in \mathbb{C}$ the product wz has modulus $|z|$ and argument $\theta + \operatorname{Arg}(z) + 2k\pi$. This means that the multiplication by w rotates z by θ counterclockwise.

Definition 8. The notation $\operatorname{cis}\theta$ represents the complex number with unit modulus and argument θ; thus, $\operatorname{cis}\theta := \cos\theta + i\sin\theta$. □

Exercises

33. Calculate the modulus and the principal argument of the complex number $(-\sqrt{3},-1)$.

34. Calculate the modulus and the principal argument of the complex number $(1,1)$.

35. Calculate the Cartesian coordinates of the complex number z with $|z| = 5$ and $\operatorname{Arg}(z) = \pi/3$.

36. Calculate the Cartesian coordinates of the complex number z with $|z| = 1$ and $\text{Arg}\,(z) = \pi/6$.

4.3 Roots

The polar coordinates give a convenient means to compute complex square roots, cube roots, and roots of every order, through trigonometry and the two formulae derived in the preceding subsection:

$$\begin{cases} |w\,z| &= |w|\,|z|, \\ \text{Arg}\,(w\,z) &= \text{Arg}\,(w) + \text{Arg}\,(z) + 2k\pi. \end{cases}$$

Indeed, if $w = z$, then the formulae just listed become

$$\begin{cases} |w^2| &= |w|^2, \\ \text{Arg}\,(w^2) &= 2\text{Arg}\,(w) + 2k\pi. \end{cases}$$

Consequently, to compute a complex square root w of a complex number z, which amounts to finding $w \in \mathbb{C}$ such that $w^2 = z$, substitute z for w^2 in the preceding formulae and solve for w:

$$\begin{cases} |z| &= |w|^2, \\ \text{Arg}\,(z) &= 2\text{Arg}\,(w) + 2k\pi; \end{cases}$$

$$\begin{cases} \sqrt{|z|} &= |w|, \\ (\text{Arg}\,(z)/2) - k\pi &= \text{Arg}\,(w). \end{cases}$$

Such a procedure yields two complex square roots. For the first root, $|w| = \sqrt{|z|}$ and $\text{Arg}\,(w) = \text{Arg}\,(z)/2$. For the second root, $|w| = \sqrt{|z|}$ and $\text{Arg}\,(w) = \text{Arg}\,(z)/2 \pm \pi$, with the sign chosen so that $\text{Arg}\,(w)$ lies in the interval $]-\pi, \pi]$, between $-\pi$ excluded and π included. To get the Cartesian coordinates of w, use the expression $w = (|w|\cos(\text{Arg}\,(w)),\ |w|\sin(\text{Arg}\,(w)))$.

Example 19. If $z = -1 = (-1, 0)$, then $|z| = 1$ and $\text{Arg}\,(z) = \pi$. For the first root, $|w| = \sqrt{1} = 1$ and $\text{Arg}\,(w) = \pi/2$, which corresponds to $w = (0, 1) = i$. For the second root, $\text{Arg}\,(w) = \pi/2 + \pi + 2k\pi = 3\pi/2 + 2k\pi = -\pi/2$ with $k = -1$ to bring the argument in the interval $]-\pi, \pi]$, and $w = (0, -1) = -i$. Verify that $i^2 = 1 = (-i)^2$, as done in **Example 13.** □

Example 20. If $z = i$, then $|z| = 1$ and $\text{Arg}\,(z) = \pi/2$. For the first root, $|w| = \sqrt{1} = 1$ and $\text{Arg}\,(w) = \pi/4$; from $\cos(\pi/4) = 1/\sqrt{2}$ and $\sin(\pi/4) = 1/\sqrt{2}$ follows $w = \left(1/\sqrt{2}, 1/\sqrt{2}\right)$. For the second root, $w = -\left(1/\sqrt{2}, 1/\sqrt{2}\right)$. Verify that $\left(1/\sqrt{2}, 1/\sqrt{2}\right)^2 = i = \left\{\left(1/\sqrt{2}, 1/\sqrt{2}\right)\right\}^2$. □

For complex roots of order greater than two, mathematical induction gives

$$\begin{cases} |w^n| &= |w|^n, \\ \operatorname{Arg}(w^n) &= n\operatorname{Arg}(w) + 2k\pi. \end{cases}$$

Hence, if $w^n = z$, then

$$\begin{cases} |z| &= |w|^n, \\ \operatorname{Arg}(z) &= n\operatorname{Arg}(w) + 2k\pi; \end{cases}$$

$$\begin{cases} \sqrt[n]{|z|} &= |w|, \\ (\operatorname{Arg}(z)/n) - 2k\pi/n &= \operatorname{Arg}(w). \end{cases}$$

The factor $2k\pi/n$ gives n roots, one for each $k \in \{0, \dots, n-1\}$.

Example 21. The cube roots of $z = i$ have modulus $|w| = \sqrt[3]{1} = 1$ and arguments $\operatorname{Arg}(w) = (\pi/2)/3 - 2k\pi/3 \in \{\pi/6, -\pi/2, 5\pi/6\}$, corresponding to the three cube roots

$$\begin{cases} (\cos(\pi/6),\ \sin(\pi/6)) &= (\sqrt{3}/2,\ 1/2), \\ (\cos(-\pi/2),\ \sin(-\pi/2)) &= (0, -1) = -i, \\ (\cos(5\pi/6),\ \sin(5\pi/6)) &= (-\sqrt{3}/2,\ 1/2). \end{cases}$$ □

Example 22. The cube roots of $z = 1$ have modulus $|w| = \sqrt[3]{1} = 1$ and arguments $\operatorname{Arg}(w) = 0/3 - 2k\pi/3 \in \{0, -2\pi/3, 2\pi/3\}$, corresponding to the three cube roots

$$\begin{cases} (\cos(0),\ \sin(0)) &= (1, 0) = 1, \\ (\cos(-2\pi/3),\ \sin(-2\pi/3)) &= (-1/2, -\sqrt{3}/2), \\ (\cos(2\pi/3),\ \sin(2\pi/3)) &= (-1/2, \sqrt{3}/2). \end{cases}$$

Observe that the three cube roots of unity are powers of one of them, called a *primitive root of unity*; for example, let $\omega := (-1/2, \sqrt{3}/2)$: then indeed, $(-1/2, -\sqrt{3}/2) = \omega^2$ and $(1, 0) = \omega^3$. Also, from $(1, 0) = \omega^3$ follow $\omega^{-1} = \omega^2$ and $(\omega^2)^{-1} = \omega$. □

Remark 6. The three cube roots of unity also serve to express all three cube roots of a nonzero complex number z in terms of any particular cube root $h = \sqrt[3]{z}$; the three cube roots of z are then h, ωh, and $\omega^2 h$. From $h^3 = z$, $\omega^3 = 1$, and $(\omega^2)^3 = 1$, it follows that

$$h^3 = z, \quad (\omega h)^3 = \omega^3 h^3 = 1z = z, \quad (\omega^2 h)^3 = (\omega^2)^3 h^3 = 1z = z.$$

As shown in the exercises, z has no cube roots other than h, ωh, and $\omega^2 h$. □

For complex square roots, algebraic formulae exist that do not involve any angle or trigonometry, as explained in the next section.

Exercises

37. Calculate all the complex cube roots of -8.

38. Calculate all the complex fourth roots of 1.

39. Prove that each nonzero complex number has exactly two complex square roots.

40. Prove that for each nonzero complex number z with cube root $h = \sqrt[3]{z}$, z has exactly three distinct cube roots, which are h, ωh, and $\omega^2 h$, with ω denoting a primitive cube root of unity.

4.4 Dot Products, Determinants, and Applications

Besides Cartesian and polar coordinates, two further concepts related to vectors in the plane prove convenient.

Definition 9. The *dot product* of two vectors (complex numbers, points) $w = (u, v)$ and $z = (x, y)$ in the plane is the real number denoted by $w \bullet z$ or $\langle w, z \rangle$ and defined by $ux + vy$:

$$\langle w, z \rangle = w \bullet z = (u, v) \bullet (x, y) := ux + vy. \qquad \square$$

The dot product relates to the Euclidean distance as follows.

Proposition 4. *For every $w = (u, v) \in \mathbb{R}^2$, the identity $|w|^2 = \langle w, w \rangle$ holds.*

Calculate

$$\langle w, w \rangle = \langle (u, v), (u, v) \rangle = uu + vv = u^2 + v^2 = |w|^2. \qquad \square$$

Definition 10. The *determinant* of two vectors (complex numbers, points) $w = (u, v)$ and $z = (x, y)$ in the plane is the real number denoted by $\det(w, z)$ and defined by $uy - vx$:

$$\det(w, z) = \det \begin{pmatrix} u & x \\ v & y \end{pmatrix} := uy - vx. \qquad \square$$

Application 1. The algebra and geometry of complex numbers simplify the notation for rotations in the plane. For instance for each point $w \in \mathbb{R}^2 = \mathbb{C}$, the point iw has the same magnitude as that of w but lies perpendicular to w and to the left of w, which means that

- $|iw| = |w|$ (both vectors have the same magnitude),
- $\langle iw, w \rangle = 0$ (the two vectors are perpendicular to each other),
- $\det(w, iw) \geq 0$ (iw lies to the left of w).

Thus, the notation iw expresses the entire statement "a vector with the same magnitude but perpendicular and to the left." Similar simplifications hold for $(\operatorname{cis}\theta)w$ and $(-\operatorname{cis}\theta)w$, which rotate w by θ counterclockwise and clockwise respectively. □

Application 2. The algebra and geometry of complex numbers simplify the notation for the circulation and expansion of a planar fluid flow with respect to any curve, for instance, a straight line. To this end, let

- $w = (u, v)$ denote the velocity of a fluid,
- $m = \operatorname{cis}\theta$ denote a unit vector parallel to the line, and
- $n = -im = -i\operatorname{cis}\theta$ denote a unit vector perpendicular to the line, to the right of m.

Then

- the real dot product $\langle w, m\rangle$ is the *circulation* of the fluid in the direction of the line, which measures the flow along the line, and
- the real dot product $\langle w, n\rangle$ is the *expansion* of the fluid in the direction perpendicular to the line, which measures the flow across the line.

Then $\langle w, m\rangle + i\langle w, n\rangle = \overline{w}m$, so that the complex multiplication $\overline{w}m$ expresses concisely the circulation (real part) and the expansion (imaginary part) [Simmonds 1996]. □

Exercises

41. Verify that the dot product is commutative and *multilinear*, which means that for all *real* numbers r_1, r_2, s_1, and s_2, and for all vectors w_1, w_2, z_1, and z_2,

$$\begin{aligned}\langle w_1, z_1\rangle &= \langle z_1, w_1\rangle,\\ \langle r_1w_1 + s_1z_1, r_2w_2 + s_2z_2\rangle &= r_1r_2\langle w_1, w_2\rangle + r_1s_2\langle w_1, z_2\rangle + s_1r_2\langle z_1, w_2\rangle\\ &\quad + s_1s_2\langle z_1, z_2\rangle.\end{aligned}$$

42. With $\angle(w, z)$ denoting the angle between w and z, use the law of cosines from plane geometry to verify that for all vectors $w, z, \in \mathbb{R}^2$,

$$\langle w, z\rangle = |w| \cdot |z| \cdot \cos(\angle(w, z)).$$

43. For all vectors $w, z, \in \mathbb{R}^2$, verify the polar identity

$$\langle w, z\rangle = (1/4)\left(|w + z|^2 - |w - z|^2\right).$$

44. For all vectors $w, z, \in \mathbb{R}^2$, verify that $|\det(w, z)|$ equals the area of the parallelogram with vertices at 0, w, z, and $w + z$.

45. Prove that if a differentiable curve $\gamma : [0, 1] \to \mathbb{C}$ parametrizes counterclockwise the boundary $\partial\Omega$ of an open set $\Omega \subset \mathbb{C}$, and if $A(\Omega)$ denotes the area of Ω, then

$$A(\Omega) = \frac{1}{2i} \oint_{\partial\Omega} \overline{z}\, dz,$$

with the integral defined by

$$\oint_{\partial\Omega} \overline{z}\, dz := \int_0^1 \overline{\gamma(t)} \cdot \gamma'(t)\, dt.$$

46. With the arithmetic of quaternions defined as in **Exercise 19**, verify that if

$$\begin{aligned} g &= (\cos(\theta/2), u\sin(\theta/2), v\sin(\theta/2), w\sin(\theta/2)), \\ 1 &= u^2 + v^2 + w^2, \\ q &= (0, x, y, z), \end{aligned}$$

then the product of quaternions $p = gq$ rotates the vector (x, y, z) in space $\mathbb{R}^3$ by θ about the line in the direction of the unit vector (u, v, w).

5. Complex Square Roots and Quadratic Equations

This section explains some of the theory that led to the complex numbers: the algebraic solution of polynomial equations. To understand the type of thoughts that enabled the discovery of the algebraic formulae to solve the cubic and quartic equations, a brief review of the solution of quadratic equations appears relevant, because complex quadratic equations form a part of the formulae to solve cubic and quartic equations, and because some of the historical methods to solve quadratic equations give some intuition into how to derive formulae for the solutions of the cubic and quartic equations.

5.1 Factoring and Solving Real Quadratics

This subsection develops various methods to solve real quadratics, which will arise in the calculation of complex square roots.

To this end, let a, b, and c represent three real numbers such that $a \neq 0$, and consider the quadratic equation

$$az^2 + bz + c = 0. \qquad \textbf{(QE)}$$

After division of both sides by a, a first method to solve the general quadratic equation **(QE)** consists of "completing the square," under the guidance of a corresponding geometric construction, which suggests adding the quantity $(b/2a)^2$ to both sides to reveal an algebraic square:

$$
\begin{aligned}
z^2 + \frac{b}{a}z + \frac{c}{a} &= 0, \\
z^2 + 2\frac{b}{2a}z + \left(\frac{b}{2a}\right)^2 + \frac{c}{a} &= \left(\frac{b}{2a}\right)^2, \\
\left(z + \frac{b}{2a}\right)^2 + \frac{c}{a} &= \left(\frac{b}{2a}\right)^2, \\
\left(z + \frac{b}{2a}\right)^2 &= \left(\frac{b}{2a}\right)^2 - \frac{c}{a}, \\
z + \frac{b}{2a} &= \pm\sqrt{\left(\frac{b}{2a}\right)^2 - \frac{c}{a}}, \\
z &= -\frac{b}{2a} \pm \sqrt{\left(\frac{b}{2a}\right)^2 - \frac{c}{a}} \\
&= -\frac{b}{2a} \pm \sqrt{\left(\frac{1}{2a}\right)^2 (b^2 - 4ac)} \\
&= \frac{-b \pm \sqrt{b^2 - 4ac}}{2a},
\end{aligned}
$$

$$
z_1 = \frac{-b + \sqrt{b^2 - 4ac}}{2a}, \quad z_2 = \frac{-b - \sqrt{b^2 - 4ac}}{2a}. \qquad \textbf{(QF)}
$$

This formula—called the *quadratic formula*—proves convenient for calculations by hand because it allows for the reading of the square root off tables and then requires only a division by 2 by hand. Similar computational advantages by hand also explain the requirements in some contexts for the rationalization of the denominator of every fraction. For automated computations, however, the formula just obtained exhibits an unacceptably large sensitivity to the rounding to a fixed finite number of digits during the intermediate computations (b^2, $4ac$, b^2-4ac, $\sqrt{b^2 - 4ac}$, and $-b\pm\sqrt{b^2 - 4ac}$). Alternative quadratic formulae have existed at least since Gulio Carlo di Fagnano's work [Fagnano 1750, 421, eq. 14], of which a particular set, displayed below, has proved better suited to finite arithmetic at least since Muller [1956], and reproduced in Forsythe [1970]. Such formulae result from a rationalization of the *numerator* of the quadratic formula.

$$x_i = \frac{-2c}{b + \operatorname{sign}(b)\sqrt{b^2 - 4ac}}$$

$$x_{ii} = \frac{b + \operatorname{sign}(b)\sqrt{b^2 - 4ac}}{-2a}$$

(AF)

The function $\operatorname{sign} : \mathbb{R} \to \mathbb{R}$ is defined by

$$\operatorname{sign}(t) := \begin{cases} 1, & \text{if } 0 \leq t; \\ -1, & \text{if } t < 0. \end{cases}$$

Equation **(QF)** already appeared in words, rather than symbols, over two thousand years ago in Babylon and in China [van der Waerden 1983, 61, 56]. Other derivations of the quadratic formula exist. Indeed, the same formula arises from a second method, based on relations between the roots, z_1 and z_2, and the coefficients, a, b, and c:

$$z^2 + \frac{b}{a}z + \frac{c}{a} = (z - z_1)(z - z_2) = z^2 - (z_1 + z_2)z + z_1 z_2.$$

Hence, equating the coefficients on the left to the corresponding coefficients on the right gives

$$\begin{cases} z_1 + z_2 &= -b/a, \\ z_1 z_2 &= c/a, \end{cases} \qquad \textbf{(QS)}$$

which thus replaces the initial quadratic equation **(QE)** by a quadratic system **(QS)**, which gives the sum and the product of the two unknowns, z_1 and z_2. To simplify the notation, let $u := z_1$, $v := z_2$, $s := -b/a$, and $r := c/a$. Then the system **(QS)** becomes

$$\begin{cases} u + v &= s, \\ uv &= r, \end{cases} \qquad \textbf{(SP)}$$

known as a "system with sum and product," and also frequent in antiquity [van der Waerden 1983, 62–63]. As in the method of completing the square, a geometric figure suggests the introduction of an auxiliary quantity d, which represents the deviation of u and v from their average $s/2$. Specifically, $d := u - s/2$, so that

$$\begin{aligned} u &= (s/2) + d, \\ v &= (s/2) - d, \\ r &= uv = ([s/2] + d)([s/2] - d) = (s/2)^2 - d^2. \end{aligned}$$

From the last equation emerges the "purely quadratic" equation $d^2 - (s/2)^2 = r$, whence

$$\begin{aligned} d &= \sqrt{(s/2)^2 - r}, \\ u &= (s/2) + d = (s/2) + \sqrt{(s/2)^2 - r}, \\ v &= (s/2) - d = (s/2) - \sqrt{(s/2)^2 - r}, \end{aligned}$$

which, after reverting to $s = -b/a$ and $r = c/a$, reduces to the quadratic formula **(QF)**.

Exercises

47. Solve $2000z^2 + z - 2 \cdot 10^{-15} = 0$ with the usual quadratic formula **(QF)**, and then again with the alternative quadratic formula **(AF)**.

48. Solve $1000z^2 + z - 10^{-15} = 0$ with the usual quadratic formula **(QF)**, and then again with the alternative quadratic formula **(AF)**.

5.2 Complex Square Roots by Algebra

This subsection explains how to establish *algebraic* formulae for the square roots of complex numbers, without trigonometry, but with algebraic methods of historical and computational interest. To establish an algebraic formula for the square root of a complex number, first write down the mathematical significance of such a square root: if $z = (x, y)$ lies in the complex plane, and if $w = (u, v)$ represents a complex square root of z, then $w^2 = z$, which leads to the following equations for u and v:

$$\begin{aligned} w^2 &= z, \\ (u, v)^2 &= (x, y), \\ (u, v)(u, v) &= (x, y), \\ (u^2 - v^2, 2uv) &= (x, y). \end{aligned}$$

Equating the real parts and the imaginary parts separately gives the following system of two equations:

$$\begin{cases} u^2 - v^2 &= x, \\ 2uv &= y. \end{cases}$$

Hence, squaring each side of the second equation leads to the system

$$\begin{cases} u^2 - v^2 &= x, \\ u^2 v^2 &= (y/2)^2, \end{cases}$$

which specifies the "difference and product" of the two unknown quantities u^2 and v^2. Such systems have been common since antiquity [van der Waerden 1983, 62–63]. Several methods exist to solve such a system, for instance, through the identity $(u^2+v^2)^2 = (u^2-v^2)^2 + 4u^2v^2$ [Ahlfors 1979, 3], or by solving the second equation for v and substituting the result in the first equation [Boas 1987, 10 and 264], or by stating and solving the problem in polar coordinates [Fisher 1986, 13], or by extracting algebraic expressions from the polar form [Churchill and Brown 1984, 21]. One of the historical methods to solve such a system, which also arises in the solution of the general cubic equation, consists of introducing the average of the two unknowns, $a := (u^2+v^2)/2$, so that $u^2 = (x/2) + a$ and $v^2 = -(x/2) + a$, whence $u^2 - v^2 = x$ and

$$a^2 - (x/2)^2 = (a + (x/2))\,(a - (x/2)) = u^2v^2 = (y/2)^2,$$

which yields $a^2 = (x/2)^2 + (y/2)^2$, and a *real* square root gives $a = \sqrt{x^2+y^2}/2$. Consequently, from $u^2 = (x/2) + a \geq 0$, another real square root leads to $u = \pm\sqrt{(x/2)+a}$. From $u^2 - v^2 = x$ it then follows that $v^2 = u^2 - x \geq 0$, so that $v = \pm\sqrt{u^2 - x}$; the equation $2uv = y$ then restricts the choice of signs, for instance

$$\sqrt{x+iy} = u(x,y) + iv(x,y),$$

$$u(x,y) = \sqrt{\frac{\sqrt{x^2+y^2}+x}{2}},$$

$$v(x,y) = \text{sign}(y)\sqrt{\frac{\sqrt{x^2+y^2}-x}{2}}.$$

Yet another method of solving for u and v, algebraically equivalent to the foregoing methods but better suited to automated digital computations, consists in deriving from the system for difference and product an equivalent quadratic equation. Indeed, the system shows that u^2 and $-v^2$ are the solutions r_i and r_{ii} of the real quadratic

$$r^2 - xr - (y/2)^2 = 0.$$

Because $u^2 \geq 0$ and $-v^2 \leq 0$, the proper choice of roots follows:

$$u^2 = \frac{x + \sqrt{x^2+y^2}}{2}, \qquad -v^2 = \frac{x - \sqrt{x^2+y^2}}{2}.$$

Hence follow two different solutions for (u,v). Fortunately, to ensure consistency, mathematicians and computer scientists have adopted the convention to choose $u \geq 0$ [Hewlett-Packard 1984; Kahan 1987]. Thus, with u unambiguously defined, return to the equation for the product, $uv = y/2$, and solve for $v = y/(2u)$.

For automated digital computations, the alternative quadratic formula gives **Algorithm 1.**

Algorithm 1. Computation of the complex square root $(u, v) = \sqrt{(x, y)}$.

$$g := |x| + \sqrt{x^2 + y^2},$$
$$g := \sqrt{g/2}.$$
If $x \geq 0$, then
$$u := g,$$
$$v := y/2u;$$
if $x < 0$, then
$$v := g \operatorname{sign}(y)$$
$$u := y/2v.$$

Example 23. Consider the particular case $z := (3, 4)$, with $x = 3$ and $y = 4$:

$$g := |3| + \sqrt{3^2 + 4^2} = 3 + 5 = 8,$$
$$g := \sqrt{g/2} = \sqrt{8/2} = 2;$$
$x \geq 0$, hence
$$u := g = 2,$$
$$v := y/2u = 4/(2 \cdot 2) = 1,$$
$$(u, v) = (2, 1);$$
$$\sqrt{(3, 4)} = (2, 1).$$

As a verification, square the result just obtained:

$$(2, 1)^2 = (2, 1)(2, 1) = (4 - 1, 2 + 2) = (3, 4). \quad ✓$$ □

Definition 11. (The Principal Complex Square Root.) The formulae just established for $(u, v) = \sqrt{(x, y)}$ constitute the *principal branch* of the complex square root function, in conformity to the mathematics literature and to computer hardware and software [Hewlett-Packard 1984, 69; Kahan 1987]. For the principal branch, the formulae reveal that v may be negative, zero, or positive, but that in all cases $u \geq 0$ by the choice of nonnegative real square roots. The adjective "principal" does not connote any greater importance: it merely specifies the choice of the complex square root, with $u \geq 0$, just as does the usual convention to interpret the symbol $\sqrt{r}$ as the positive square root, rather than the negative square root, of a positive real number r. □

Remark 7. *(Other square roots.)* Each complex number (x, y) has two complex square roots: (u, v) and $-(u, v)$. While v and $-v$ may assume any real value, $u \geq 0$ in the principal branch (u, v), and $-u \leq 0$ in the other branch, $-(u, v) = (-u, -v)$. □

Example 24. Let $z := i = (0, 1)$. Then $x = 0$ and $y = 1$. Consequently,

$$u = \sqrt{(0 + \sqrt{0^2 + 1^2})/2} = 1/\sqrt{2}, \qquad v = 1 \cdot \sqrt{(-0 + \sqrt{0^2 + 1^2})/2} = 1/\sqrt{2},$$

whence the principal branch of the square root becomes $(u, v) = (1/\sqrt{2}, 1/\sqrt{2}) = (1, 1)/\sqrt{2}$:

$$\sqrt{i} = \left(\frac{1}{\sqrt{2}}, \frac{1}{\sqrt{2}}\right).$$

As a verification, square both square roots:

$$\left(\pm\sqrt{i}\right)^2 = \left(\frac{1}{\sqrt{2}}, \frac{1}{\sqrt{2}}\right)^2 = \cdots = (0, 1). \quad \checkmark$$

□

Application 3. Interestingly, the principal branch of the complex square root $\sqrt{\ } : (\mathbb{C} \setminus \mathbb{R}_-) \to \mathbb{C}$ helps in the numerical computations of trajectories with near-collisions [Hut 1985; Stiefel and Scheifele 1971]. In such problems, an object (particle, satellite, star) moves close to the negative real axis towards another object (particle, planet, star) at the origin. For very close encounters, computers' finite arithmetic may fail to identify the outcome (collision, capture, escape). In such situations, the principal branch of the complex square root transforms the original trajectory with its narrow vertex near the origin into a nearly straight curve in the right-hand half-plane, which facilitates computations. □

Exercises

49. Calculate the following principal square roots.

a) $\sqrt{(5, 12)}$ **b)** $\sqrt{(7, 24)}$

50. Calculate the following principal square roots.

a) $\sqrt{(15, 8)}$ **b)** $\sqrt{(35, 12)}$

51. In **Example 23** on p. 30, and in the preceding exercises, the complex number (h, k) has integer coordinates h and k and integer modulus $\ell := |(h, k)| = \sqrt{h^2 + k^2}$. In such a situation, (h, k, ℓ) is called a *Pythagorean triple*. Also, in **Example 23**, and in the preceding exercises, $\sqrt{(h, k)}$ has integer coordinates. Determine whether $\sqrt{(h, k)}$ has integer coordinates for every Pythagorean triple (h, k, ℓ).

52. Let $\mathcal{R}(\sqrt{\ })$ represent the graph of the principal branch of the complex square root function:

$$\mathcal{R}(\sqrt{\ }) := \{(z, w) \in \mathbb{C}^2 \,:\, w^2 = z\}.$$

Find a bijection $f : \mathbb{C} \to S$; prove that f is a bijection. $\mathcal{R}(\sqrt{\ })$ is also called the *Riemann surface* of $\sqrt{\ }$.

53. This exercise gives another complex square root function, denoted here by $S(x, y) = (p, q)$, which is analogous, but not identical, to the principal branch. To this end, return to the specification that $(p, q)^2 = (x, y)$, and solve the system for the difference and product first for q. Show that the resulting formulae for p and q are analogous, but not identical, to those for u and v.

54. (†) In contrast to the complex square root, prove that the complex cube root, $\sqrt[3]{z} = \sqrt[3]{(x, y)}$, does *not* admit any algebraic formula involving only real arithmetic and real roots of any order in terms of the real numbers x and y.

55. With $\sqrt{\ }$ denoting the principal branch of the complex square root, determine whether the following equality holds for every pair of complex numbers w and z, and justify your answer:

$$\sqrt{w\,z} = \sqrt{w}\,\sqrt{z}.$$

56. Determine whether the following equality holds for every nonzero complex number z, and prove your answer:

$$\frac{z}{|z|} = \big(\cos(\text{Arg}\,(z)),\ \sin(\text{Arg}\,(z))\big).$$

The following exercises pertain to the principal branch of the complex square root $\sqrt{\ } : (\mathbb{C} \setminus \mathbb{R}_-) \to \mathbb{C}$.

57. **a)** Identify the images by $\sqrt{\ }$ of horizontal straight lines (parallel to the first coordinate axis).

b) Identify the images by $\sqrt{\ }$ of vertical straight lines (parallel to the second coordinate axis).

c) Identify the images by $\sqrt{\ }$ of straight lines passing through the origin.

d) Identify the images by $\sqrt{\ }$ of straight lines in the plane.

58. Verify that the principal branch of the complex square root maps the complex plane cut along the nonpositive axis, $\mathbb{C} \setminus \mathbb{R}_-$, onto the right-hand half-plane, $\{z \in \mathbb{C} : \text{Re}(z) > 0\}$.

59. Identify the images by $\sqrt{\ }$ of arcs of circles centered at the origin.

60. a) Identify the image by $\sqrt{}$ of a fictitious trajectory consisting of an arc of a circle centered at the origin connected to two horizontal half-lines parallel to the negative real axis.

b) Identify the image by $\sqrt{}$ of a real parabolic trajectory with axis along the negative real axis and focus at the origin.

5.3 Complex Quadratic Equations

Let a, b, and c represent three complex numbers such that $a \neq 0$, and consider the quadratic equation

$$az^2 + bz + c = 0. \tag{QE}$$

After division of both sides by a, a first method to solve the general quadratic equation **(QE)** consists of "completing the square," under the guidance of a corresponding geometric construction, which suggests adding the quantity $(b/[2a])^2$ to both sides to reveal an algebraic square, *verbatim* as in the derivation of the real quadratic formula **Q**:

$$z_1 = \frac{-b + \sqrt{b^2 - 4ac}}{2a}, \quad z_2 = \frac{-b - \sqrt{b^2 - 4ac}}{2a}. \tag{QF}$$

Example 25. To solve $z^2 + iz - (1 + i) = 0$, observe that $a = 1 = (1, 0)$, $b = i = (0, 1)$, and $c = -(1 + i) = (-1, -1)$. Hence, calculate

$$b^2 - 4ac = (0, 1)^2 - 4(1)(-1, -1) = (-1, 0) + (4, 4) = (3, 4).$$

Also, **Example 23** showed that $\sqrt{(3, 4)} = (2, 1)$. Consequently,

$$z_1 = \frac{-b + \sqrt{b^2 - 4ac}}{2a} = \frac{-(0, 1) + (2, 1)}{2(1, 0)} = \frac{(2, 0)}{2} = 1,$$

$$z_2 = \frac{-b - \sqrt{b^2 - 4ac}}{2a} = \frac{-(0, 1) - (2, 1)}{2(1, 0)} = \frac{(-2, -2)}{2} = -(1, 1).$$

Substitutions into the given equation confirm the results just obtained. □

Remark 8. For digital computations, the alternative quadratic formula of **Algorithm 2** proves less sensitive to rounding than the usual quadratic formula [Hammerlin and Hoffmann 1991, 17; Stoer and Bulirsch 1983, 20–22]. Such an alternative formula first selects the root among z_1 and z_2 that is less sensitive to rounding errors, denoted here by z_I, and then computes the other root as $z_{II} = (c/a)/z_I$. □

Algorithm 2. Numerical Solution of $az^2 + bz + c = 0$ for $a, b, c \in \mathbb{C}$ with $a \neq 0$

- *Step 0.* If $b = 0$ then $z = \pm\sqrt{-c/a}$. If $b \neq 0$, then proceed as follows.
- *Step 1.* Compute $p := 2a$ and $q := 2c$.
- *Step 2.* Compute $s := \sqrt{b^2 - pq}$ by **Algorithm 1**; thus $s = \sqrt{b^2 - 4ac}$.
- *Step 3.* Compute the *dot* product $r := \langle b, s \rangle = b_1 s_1 + b_2 s_2$, with $b = (b_1, b_2)$ and $s = (s_1, s_2)$.
- *Step 4.* If $r \geq 0$, then compute $t := -(b + s)$. If $r < 0$, then compute $t := s - b$.
- *Step 5.* Compute the two roots by the formulae $z_I := t/p$ and $z_{II} := q/t$.

Exercises

61. Solve the quadratic equation $(1, 3)z^2 + z - 1 = 0$.

62. Solve the quadratic equation $z^2 - 2z + 2 = 0$.

6. Equations Solvable by Radicals

6.1 Complex Cubic Equations

Let c_0, c_1, c_2, and c_3 represent four complex numbers such that $c_3 \neq 0$, and consider the cubic equation

$$c_3 z^3 + c_2 z^2 + c_1 z + c_0 = 0. \tag{CE}$$

In contrast with quadratic equations, no affine change of variable of the type $z := w - t$ can "complete the cube," as proved in the exercises. Nevertheless, the translation by $t := c_2/(3c_3)$ and the substitution $z := w - t$ annihilates the quadratic term and leads to the "reduced cubic equation"

$$w^3 + pw + q = 0, \tag{RC}$$

with p and q computable efficiently with an auxiliary quantity f as follows:

$$t := \frac{c_2}{3c_3}, \qquad f := \frac{c_1}{c_3},$$
$$q := \frac{c_0}{c_3} + \left(2t^2 - f\right) \cdot t, \qquad p := f - 3t^2.$$

Hence, inspired perhaps by systems with sums and products **(SP)**, or perhaps by the sum and difference in the quadratic formula **(QF)**, Cardano solved the reduced cubic equation through the substitution $w := u + v$, which gives

$$\begin{aligned} w^3 + pw + q &= 0, \\ (u+v)^3 + p(u+v) + q &= 0, \\ u^3 + 3u^2v + 3uv^2 + v^3 + p(u+v) + q &= 0, \\ u^3 + v^3 + 3uv(u+v) + p(u+v) + q &= 0, \end{aligned}$$

and which would hold if u and v satisfied the system with sum and product

$$\begin{cases} u^3 + v^3 &= -q, \\ 3uv &= -p; \end{cases}$$

whence

$$\begin{cases} u^3 + v^3 &= -q, \\ u^3v^3 &= (-p/3)^3. \end{cases}$$

Thus, u^3 and v^3 are the solutions r_i and r_{ii} of the quadratic equation

$$r^2 + qr - (p/3)^3 = 0.$$

Whether $u^3 = r_i$ or $u^3 = r_{ii}$ does not matter for the purpose of computing $w = u + v$. Hence, the alternative complex quadratic formula gives u^3,

$$\begin{aligned} d &:= \sqrt{(q/2)^2 + (p/3)^3}, & r &:= \langle q, d \rangle, \\ s &:= \operatorname{sign}(r), & u^3 &:= -[(q/2) + sd], \end{aligned}$$

whence the product equation $3uv = -p$ and the three complex cube roots, indexed by $k \in \{0, 1, 2\}$, produce the three solutions

$$\begin{aligned} u_k &= \sqrt[3]{u^3}, \\ v_k &= -p/(3u_k), \\ w_k &= u_k + v_k, \\ z_k &= w_k - t. \end{aligned}$$

Alternatively, solving the system just obtained for u^3 and v^3 by the method developed previously yields

$$\begin{aligned} u^3 &= (-q/2) + d, \\ v^3 &= (-q/2) - d, \\ (-p/3)^3 = u^3v^3 &= (-q/2)^2 - d^2. \end{aligned}$$

Hence, solve for d, substitute the result for $u^3 = (-q/2) + d$, compute one value of u_k, $k \in \{0, 1, 2\}$, for each complex cube root, and then use the given product $u_k v_k = -p/3$ to get $v_k = -p/(3u_k)$ to get all three roots:

$$\begin{aligned} d &= \sqrt{(q/2)^2 + (p/3)^3}, \\ u^3 &= (-q/2) + d, \\ u_k &= \sqrt[3]{d - (q/2)}, \\ v_k &= -p/(3u_k), \\ w_k &= u_k + v_k, \\ z_k &= w_k - t. \end{aligned}$$

Two particular cases deserve mention. If $u = 0$ but $v \neq 0$, then solve first for $v^3 = (-q/2) - d$ and then set $u = -p/(3v)$. If $u = 0 = v$, then elementary algebra shows that $p = 0 = q$, whence the reduced cubic equation **(RC)** has one triple root, $w = 0$.

Example 26. The cubic equation $z^3 - 6z^2 + 11z - 6 = 0$ has coefficients $c_3 = 1$, $c_2 = -6$, $c_1 = 11$, and $c_0 = -6$, which lead to the values $t = -2$, $f = 11$, $q = 0$, $p = -1$, and

$$u^3 = -(q/2) + \text{sign}(q)\sqrt{(-q/2)^2 + (p/3)^3} = 0 + \sqrt{-1/27} = i/(3\sqrt{3}).$$

From **Example 21** on p. 22 follows $\sqrt[3]{i} = -i$. Consequently,

$$\begin{aligned} u_0 &= \sqrt[3]{u^3} = \sqrt[3]{i/(3\sqrt{3})} = -i/\sqrt{3}, \\ v_0 &= -p/(3u_0) = 1/(-3i/\sqrt{3}) = i/\sqrt{3}, \\ w_0 &= u_0 + v_0 = -i/\sqrt{3} + i/\sqrt{3} = 0, \\ z_0 &= w_0 - t = 0 - (-2) = 2. \end{aligned}$$

Moreover, from **Remark 6** on p 22, where $\omega = (-1/2, -\sqrt{3}/2)$, follow the other two roots $u_1 = \omega u_0$ and $u_2 = \omega^2 u_0$, whence arithmetic similar to that for z_0 gives $z_1 = w_1 + 2 = 1 + 2 = 3$ and $z_2 = w_2 + 2 = -1 + 2 = 1$. Substitutions into the given equation confirm the three solutions 1, 2, and 3. □

Remark 9. Another method, due to Vieta, also begins by reducing the initial cubic equation, but then consists of performing the substitution $w := s - p/(3s)$ in the reduced cubic $w^3 + pw + q = 0$, and in multiplying the resulting equation by s^3, which gives $s^6 - (p/3)^3 + qs^3 = 0$, in effect a quadratic equation in s^3. Either of the two solutions

$$s^3 = \left(-q \pm \sqrt{q^2 + 4(p/3)^3}\right)/2 = -(q/2) \pm \sqrt{(q/2)^2 + (p/3)^3}$$

gives three values of $s = \sqrt[3]{-(q/2) \pm \sqrt{(q/2)^2 + (p/3)^3}}$ and hence three values of $w = s - p/(3s)$. See also Birkhoff and Mac Lane [1977, 119] and

Dickson [1914, 31–32]. Though Vieta's derivation seems different from Cardano's, observe that both methods lead to the same calculations, because $\sqrt{(q/2)^2 + (p/3)^3} = d$, $s = u$, $v = -p(3s)$, and $w = s - p/(3s) = u + v$. □

Remark 10. There are special methods for cubic equations with only rational coefficients, for instance, to investigate whether they admit rational solutions [Hungerford 1974, 160], or to determine all the solutions if at least one rational solution exists [Schulz 1982]. □

Remark 11. The numerical solution of cubic equations need not repeat the whole derivation, but may employ only the resulting formulae, for instance, as in **Algorithm 3.** □

Exercises

63. Solve the cubic equation $z^3 - 4z^2 + z + 6 = 0$.

64. Solve the cubic equation $z^3 - 4z^2 + 5z - 2 = 0$.

65. Compute the positive solution of the following cubic equation:

$$x^3 + 0.1000269x^2 - 8.070\,000\,100\,7 \times 10^{-7}x - 2.708\,83 \times 10^{-19} = 0.$$

66. In the cubic equation $c_3z^3 + c_2z^2 + c_1z + c_0 = 0$, make the substitution $z := w - t$, and verify that in the resulting equation, the coefficient of w^3 vanishes if, but only if, $t = c_2/(3c_3)$.

67. Find the dimensions—height, length, and width—of a parallelipipedic temple subject to the following specifications:

- (A) the sum of all three dimensions—height, length, and width—equals 10,
- (B) the sum of the areas of all six faces equals 62, and
- (C) the volume equals 30.

68. With u_0 and v_0 as in the text, prove that the three roots z_0, z_1, and z_2 also admit the following alternative expressions:

$$z_0 = u_0 + v_0 - t, \quad z_1 = (\omega u_0) + (\omega^2 v_0) - t, \quad z_2 = (\omega^2 u_0) + (\omega v_0) - t.$$

69. (†) Define the *discriminant* D_3 of the cubic polynomial p with

$$p(z) = c_3z^3 + c_2z^2 + c_1z + c_0$$

Algorithm 3. Numerical Solution of $c_3 z^3 + c_2 z^2 + c_1 z + c_0 = 0$ for $c_3, c_2, c_1, c_0 \in \mathbb{C}$ with $c_3 \neq 0$

- *Step 1.* Compute the following quantities, where $\widetilde{p} = -p/3$:

$$t := \frac{c_2}{3c_3,} \qquad h := t^2, \qquad f := \frac{c_1}{c_3},$$
$$q := \frac{c_0}{c_3} + (2h - f) \cdot t, \qquad \widetilde{p} := h - \frac{f}{3},$$

- *Step 2.* If $q \neq 0$ or $\widetilde{p} \neq 0$, then skip to step 3. Otherwise, if $q = 0$ and $\widetilde{p} = 0$, then the initial cubic equation has one triple root:

$$z_0 = z_1 = z_2 = -t.$$

- *Step 3.* At this point, $q \neq 0$ or $\widetilde{p} \neq 0$; compute the following quantities, where $\langle \, , \rangle$ denotes the real *dot* product:

$$\begin{aligned} &\widetilde{q} := -q/2, \\ &d := \sqrt{(\widetilde{q})^2 - (\widetilde{p})^3} \quad \text{choose either root (\textbf{Algorithm 1})}, \\ &r := \langle \widetilde{q}, d \rangle = \widetilde{q}_1 d_1 + \widetilde{q}_2 d_2. \end{aligned}$$

- *Step 4.* Compute

$$s := \begin{cases} \widetilde{q} + d, & \text{if } r \geq 0; \\ \widetilde{q} - d & \text{if } r < 0. \end{cases}$$

- *Step 5.* The condition that $q \neq 0$ or $\widetilde{p} \neq 0$ guarantees that $s \neq 0$; compute the three values

$$u_k = \sqrt[3]{s} \quad \text{(compute all three roots, } u_0 = \sqrt[3]{s} \text{ (any cube root),}$$
$$u_1 = \omega u_0, \text{ and } u_2 = \omega^2 u_0.)$$

- *Step 6.* At this point, $s \neq 0$ whence every $u_k \neq 0$; compute the roots z_0, z_1, and z_2 as follows, for $k \in \{0, 1, 2\}$:

$$v_k = \frac{\widetilde{p}}{u_k}, \qquad w_k = u_k + v_k, \qquad z_k = w_k - t.$$

as a multiple (c_3^4) of the product of the squares of the differences of pairs of distinct roots of the cubic polynomial p:

$$D_3(c_0, c_1, c_2, c_3) := c_3^4 (z_2 - z_3)^2 (z_3 - z_1)^2 (z_1 - z_2)^2.$$

a) Using any of the several expressions for the roots, verify the algebraic identity

$$\begin{aligned} D_3(c_0, c_1, c_2, c_3) &= 18c_0c_1c_2c_3 + (c_1c_2)^2 - \left[27(c_0c_3)^2 + 4c_0c_2^3 + 4c_3c_1^3\right] \\ &= -c_3^4(4p^3 + 27q^2). \end{aligned}$$

b) Prove that p has a multiple root if, but only if, $D_3(c_0, c_1, c_2, c_3) = 0$.

c) Conclude that if all the coefficients of p are real, then p has three real roots if, but only if, $D_3(c_0, c_1, c_2, c_3) > 0$.

70. Determine an algebraic condition on the coefficients of a cubic polynomial for that polynomial to have a triple root.

6.2 Trigonometric Solutions of Real Cubic Equations

For cubic polynomials with real coefficients and three real roots, an alternative method of solution exists, which avoids complex cube roots but uses real trigonometric functions instead [Dickson 1914, 34–35]. To derive such formulae, return to the equivalent system for sum and product,

$$\begin{cases} u^3 + v^3 &= -q, \\ u^3v^3 &= (-p/3)^3, \end{cases}$$

with u^3 and v^3 solutions of the quadratic equation

$$r^2 + qr - (p/3)^3 = 0,$$

for instance,

$$d := \sqrt{(q/2)^2 + (p/3)^3}, \qquad u^3 := -(q/2) + d.$$

With all the coefficients of the cubic polynomial real, and with three real roots, the discriminant shows that $(q/2)^2 + (p/3)^3 < 0$. Consequently, the polar representations of u^3 and v^3 become

$$\begin{aligned} u^3 &= -(q/2) + d = r\left[\cos\theta + i\sin\theta\right], \\ v^3 &= -(q/2) - d = r\left[\cos\theta - i\sin\theta\right], \end{aligned}$$

with

$$\begin{aligned} &r^2 = |-(q/2)|^2 + |d|^2 = |-(q/2)|^2 - [(q/2)^2 + (p/3)^3] = -(p/3)^3, \\ &\cos\theta = -q/(2r), \qquad \sin\theta = \sqrt{-[(q/2)^2 + (p/3)^3]}/r. \end{aligned}$$

Hence, the cube roots

$$u_k = \sqrt[3]{u^3} = \sqrt[3]{r}\left[\cos\left(\frac{\theta + 2k\pi}{3}\right) + i\sin\left(\frac{\theta + 2k\pi}{3}\right)\right],$$
$$v_k = \sqrt[3]{u^3} = \sqrt[3]{r}\left[\cos\left(\frac{\theta + 2k\pi}{3}\right) - i\sin\left(\frac{\theta + 2k\pi}{3}\right)\right]$$

ensure that $u_k v_k = -p/3$, and, therefore, that each $z_k = u_k + v_k - t$ solves the cubic:

$$\cos\theta = \frac{-q}{2}\sqrt{\frac{-27}{p^3}}, \qquad z_k = 2\sqrt{\frac{-p}{3}}\cos\left(\frac{\theta + 2k\pi}{3}\right) - \frac{c_2}{3c_3}, \qquad k \in \{0, 1, 2\}.$$

Example 27. The cubic equation $z^3 - 6z^2 + 11z - 6 = 0$ has real coefficients $c_3 = 1$, $c_2 = -6$, $c_1 = 11$, and $c_0 = -6$, which lead to the values $t = -2$, $f = 11$, $q = 0$, $p = -1$. Consequently,

$$(q/2)^2 + (p/3)^3 = -1/27 < 0,$$

which means that the equation has three distinct real roots. Hence,

$$\cos\theta = \frac{-q}{2}\sqrt{\frac{-27}{p^3}} = \frac{-0}{2}\sqrt{\frac{-27}{p^3}} = 0, \qquad \theta = \text{Arccos}\,(0) = \pi/2,$$
$$z_k = 2\sqrt{\frac{-p}{3}}\cos\left(\frac{\theta + 2k\pi}{3}\right) - \frac{c_2}{3c_3} = 2\sqrt{\frac{1}{3}}\cos\left(\frac{(\pi/2) + 2k\pi}{3}\right) - \frac{-6}{3},$$
$$z_0 = 2\sqrt{\frac{1}{3}}\cos\left(\frac{\pi}{6}\right) + 2 = 2\sqrt{\frac{1}{3}}\frac{\sqrt{3}}{2} + 2 = 1 + 2 = 3,$$
$$z_1 = 2\sqrt{\frac{1}{3}}\cos\left(\frac{5\pi}{6}\right) + 2 = 2\sqrt{\frac{1}{3}}\frac{-\sqrt{3}}{2} + 2 = -1 + 2 = 1,$$
$$z_2 = 2\sqrt{\frac{1}{3}}\cos\left(\frac{9\pi}{6}\right) + 2 = 2\sqrt{\frac{1}{3}} \cdot 0 + 2 = 2.$$

Application 4. Cubic equations arise in the calculation of confocal coordinates for the study of the gravitational potential of ellipsoids [Grossman 1996], as outlined in the exercises. □

Exercises

71. Solve the cubic equation $z^3 - 4z^2 + z + 6 = 0$.

72. Solve the cubic equation $z^3 - 4z^2 + 5z - 2 = 0$.

73. (†) Consider a family $\mathcal{F}_{a,b,c}$ of confocal quadric surfaces,

$$\mathcal{F}_{a,b,c} := \{S(\lambda) : \lambda \in \mathbb{R}\},$$

with each surface $S(\lambda)$ defined by the equation

$$\frac{x^2}{a^2+\lambda} + \frac{y^2}{b^2+\lambda} + \frac{z^2}{c^2+\lambda} = 1.$$

Establish an algorithm, formula, or method to compute all the values of λ satisfying the equation just displayed, in terms of (x, y, z) and (a, b, c). The resulting values of λ are called the *confocal coordinates* of the point (x, y, z) with respect to the reference ellipsoid $S(0)$, and they play a role in the study of the gravitational potential of a homogeneous ellipsoid.

74. Prove that each real cubic polynomial has at least one real root.

6.3 Complex Quartic Equations

Let c_0, c_1, c_2, c_3, and c_4 denote five complex numbers such that $c_4 \neq 0$, and consider the general quartic equation, of degree four,

$$c_4 z^4 + c_3 z^3 + c_2 z^2 + c_1 z + c_0 = 0. \qquad \textbf{(FE)}$$

The original solution, developed by Lodovico Ferrari about in 1543, proceeds as follows [van der Waerden 1985, 56–59]. As with cubic equations, a translation $z := w - t$ annihilates the cubic term in **(FE)**: a substitution of $w - t$ for z in **(FE)** shows that such a cancellation occurs for $t := c_3/(4c_4)$. Then a division by c_4 leads to the reduced quartic

$$w^4 + aw^2 + bw + c = 0, \qquad \textbf{(RQ)}$$

where

$$t = \frac{c_3}{4c_4}, \qquad a = \frac{c_2}{c_4} - \frac{3c_3^2}{8c_4^2} = \frac{c_2}{c_4} - 6t^2,$$

$$b = \frac{c_1}{c_4} - \frac{c_2c_3}{2c_4^2} + \frac{c_3^3}{8c_4^3} = \frac{c_1}{c_4} - 2\frac{c_2}{c_4}t + 8t^3,$$

$$c = \frac{c_0}{c_4} - \frac{c_1c_3}{4c_4^2} + \frac{c_2c_3^2}{16c_4^3} - \frac{3c_3^4}{256c_4^4} = \frac{c_0}{c_4} - \frac{c_1}{c_4}t + \frac{c_2}{c_4}t^2 - 3t^4,$$

Ferrari's idea consists of completing the square for the terms $w^4 + aw^2$, by adding $aw^2 + a^2$ to both sides:

$$w^4 + aw^2 + bw + c = 0,$$

$$(w^2 + a)^2 = aw^2 - bw - c + a^2.$$

Unfortunately, the right-hand side is not the square of an affine expression in w. Yet, as for the quadratic equation, a geometric figure suggests the following identity, due to Ferrari, and readily verified by algebra:

$$(w^2 + a + d)^2 = (w^2 + a)^2 + 2w^2 d + 2ad + d^2. \tag{FI}$$

Thus, substituting for $(w^2+a)^2$ the expression $aw^2 - bw - c + a^2$ just obtained yields the factored form

$$\begin{aligned}(w^2 + a + d)^2 &= (aw^2 - bw - c + a^2) + 2w^2 d + 2ad + d^2 \\ &= (a + 2d)w^2 - bw + (a^2 + 2ad + d^2 - c). \end{aligned}\tag{FF}$$

A judicious choice of d may transform the right-hand side into a square of the form $(pw + q)^2$:

$$(a + 2d)w^2 - bw + (a^2 + 2ad + d^2 - c) = (pw + q)^2 = p^2w^2 + 2pqw + q^2$$

if $(a + 2d) = p^2$, $-b = 2pq$, and $(a^2 + 2ad + d^2 - c) = q^2$, which gives the equations

$$\begin{aligned}(a + 2d)(a^2 + 2ad + d^2 - c) = p^2q^2 = (pq)^2 = (-b/2)^2, \\ 2d^3 + 5ad^2 + (4a^2 - 2c)d + a(a^2 - c) - (-b/2)^2 = 0.\end{aligned}\tag{RE}$$

Notice that equation **(RE)**, called the *resolvant cubic equation*, lends itself to Cardano's solution of the general cubic equation **(CE)** and admits three solutions, d_0, d_1, and d_2.

Hence, solving for d, and substituting any *one* solution d into the factored form **(FF)** of the reduced quartic, gives the biquadratic equation

$$(w^2 + a + d)^2 = \left(w\sqrt{a + 2d} + \sqrt{a^2 + 2ad + d^2 - c}\right)^2, \tag{BQ}$$

with the complex square roots $p = \sqrt{a + 2d}$ and $q = \sqrt{a^2 + 2ad + d^2 - c}$ chosen so that $pq = -b/2$; numerically, this choice means that if $a + 2d \neq 0$, then compute $p = \sqrt{a + 2d}$ and then $q = -b/(2p)$; if $a + 2d = 0$ but $a^2 + 2ad + d^2 - c \neq 0$, then compute $q = \sqrt{a^2 + 2ad + d^2 - c}$ and then $p = -b/(2q)$. Solving for w then involves the solutions of two quadratic equations,

$$\begin{aligned}w^2 + a + d &= pw + q, \\ w^2 + a + d &= -(pw + q),\end{aligned}$$

which give four solutions, w_k, with $k \in \{0, 1, 2, 3\}$. Finally, the translation $z_k = w_k - t$ yields the four solutions of the general quartic equation.

Remark 12. *(Factoring real quartics.)* If the initial quartic equation **(FE)** has only real coefficients, $c_0, c_1, c_2, c_3, c_4 \in \mathbb{R}$, then so does the reduced quartic. Hence, with $a, b, c \in \mathbb{R}$, the resolvant cubic equation **(RE)**, in the form

$$(a + 2d)([a + d]^2 - c) = (-b/2)^2, \tag{RE}$$

has a real solution d such that $a + 2d \geq 0$ and $(a + d)^2 - c \geq 0$. Indeed, if $a + 2d = 0$ or if $[a + d]^2 - c = 0$, then the left-hand side of **(RE)** equals zero, $(a+2d)([a+d]^2-c) = 0$; yet $(a+2d)([a+d]^2-c)$ tends to infinity as d increases; consequently, for some d large enough, $(a + 2d)([a + d]^2 - c) = (-b/2)^2$ with $a+2d \geq 0$ and $(a+d)^2-c \geq 0$. Such a solution d guarantees that both square roots in the biquadratic equation **(BQ)**, $\sqrt{a + 2d}$ and $\sqrt{a^2 + 2ad + d^2 - c}$, are real, with their product equal to $-b/2$. Therefore, **(BQ)** shows that the reduced quartic equation factors as the product of real quadratics:

$$\begin{aligned} w^4 + aw^2 + bw + c &= (w^2 + a + d)^2 - \left(w\sqrt{a + 2d} + \sqrt{[a + d]^2 - c}\right)^2 \\ &= \left[(w^2 + a + d) + \left(w\sqrt{a + 2d} + \sqrt{[a + d]^2 - c}\right)\right] \\ &\quad \cdot \left[(w^2 + a + d) - \left(w\sqrt{a + 2d} + \sqrt{[a + d]^2 - c}\right)\right] \\ &= \left[w^2 + w\sqrt{a + 2d} + a + d + \sqrt{[a + d]^2 - c}\right] \\ &\quad \cdot \left[w^2 - w\sqrt{a + 2d} + a + d - \sqrt{[a + d]^2 - c}\right]. \end{aligned}$$ □

Remark 13. *(Other methods.)* A method similar to Ferrari's also succeeds without the annihilation of the cubic term, but then two separate cases arise, according to whether $c_3^2 - 4c_2 + 4d$ equals zero [Dickson 1914, 38]. Other algebraic methods exist, due to Descartes, Euler, and Lagrange [Dickson 1914, Ch. IV, 38–46]. □

Remark 14. *(Applications.)* Quartic equations arise in the computation of the internal rate of return of such investments as Two-Year US Treasury Notes, which mature after four semi-annual compounding periods [Paley et al. 1984]. Though Ferrari's algebraic solution may provide theoretical information, or an initial estimate of which of the four roots pertains to the application at hand, practical computations may prove faster and more accurate with a numerical approximation, for instance, Newton's Method, than with lengthy algebraic formulae involving square and cube roots and inverse trigonometric functions [Macleod 1984]. Moreover, Ferrari's formulae only give an illusion of exactness: The computations of the real square roots, cube roots, and trigonometric functions for the complex cube roots all require iterative methods [Pulskamp and Delaney 1992]. □

Remark 15. The numerical solution of quartic equations need not repeat the whole derivation, but may employ only the resulting formulae, for instance, as in **Algorithm 4.** □

Algorithm 4. Numerical Solution of $c_4z^4 + c_3z^3 + c_2z^2 + c_1z + c_0 = 0$ for $c_4, c_3, c_2, c_1, c_0 \in \mathbb{C}$ with $c_4 \neq 0$

- *Step 1.* Compute the following quantities,

$$t := \frac{c_3}{4c_4}, \qquad h := \frac{c_2}{c_4}, \qquad a := h - 6t^2,$$
$$k := \frac{c_1}{c_4}, \qquad b := \left(4t^2 - 2h\right) t + k, \qquad \ell := \frac{c_0}{c_4},$$
$$c := \left([h - 3t^2]t - k\right) t + \ell.$$

- *Step 2.* Solve the following resolvant cubic for d with **Algorithm 3**:

$$2d^3 + 5ad^2 + (4a^2 - 2c)d + [a(a^2 - c) - (-b/2)^2] = 0. \qquad \textbf{(RE)}$$

- *Step 3.* Compute p and q as follows:

$$p := a + 2d,$$
$$q := (a + d)^2 - c;$$

if $|p| \geq |q|$, then

$$p := \sqrt{a + 2d}, \ \textbf{(Algorithm 1)}$$
$$q := -b/(2p);$$

whereas if $|p| < |q|$, then

$$q := \sqrt{(a + d)^2 - c}, \ \textbf{(Algorithm 1)}$$
$$p := -b/(2q).$$

The reduced quartic then factors in the form

$$w^4 + aw^2 + bw + c = \left[w^2 + w\sqrt{a + 2d} + a + d + \sqrt{[a + d]^2 - c}\right] \cdot \left[w^2 - w\sqrt{a + 2d} + a + d - \sqrt{[a + d]^2 - c}\right].$$

- *Step 4.* Solve each of the following quadratics for w with **Algorithm 2**:

$$w^2 + a + d = pw + q,$$
$$w^2 + a + d = -(pw + q).$$

- *Step 5.* Denote the four solutions by w_0 and w_1 for the first equation, and by w_2 and w_3 for the second equation. Then the four solutions of the initial quartic are

$$z_k := w_k - t, \quad k \in \{0, 1, 2, 3\}.$$

Exercises

75. Apply Ferrari's method to factor the quartic polynomial

$$z^4 - 2z^3 - 7z^2 + 8z + 12.$$

76. Apply Ferrari's method to factor the quartic polynomial

$$z^4 - 3z^3 + 2z^2 + z + 5.$$

77. (†) (Proposed by W. Weston Meyer, General Motors Research and Environmental Staff, Warren, MI.) Show that the quartic equation

$$z^4 - 2cz^3 + 2\overline{c}z - 1 = 0,$$

where c is a complex number with complex conjugate $\overline{c}$, has a root not on the unit circle $\{z : |z| = 1\}$ if and only if $(\mathrm{Re}(c))^{1/3} + i(\mathrm{Im}(c))^{1/3}$ lies outside this circle.

78. In the quartic equation $c_4z^4+c_3z^3+c_2z^2+c_1z+c_0 = 0$, make the substitution $z := w - t$, and verify that in the resulting equation, the coefficient of w^3 vanishes if, but only if, $t = c_3/(4c_4)$.

79. (†) Define the *discriminant* D_4 of the quartic polynomial q with

$$q(z) = c_4z^4 + c_3z^3 + c_2z^2 + c_1z + c_0$$

by

$$\begin{aligned} I(c_0, c_1, c_2, c_3, c_4) &:= c_4c_0 - c_3c_1/4 + 3(c_2/6)^2, \\ I(c_0, c_1, c_2, c_3, c_4) &:= c_4c_2c_0/6 + c_3c_2c_1/48 - c_3c_1^2/64 - (c_2/6)^3 - c_0(c_4/4)^2, \\ D_4 &:= I^3 - 27J^2. \end{aligned}$$

Prove that q has a multiple root if, but only if, $D_4(c_0, c_1, c_2, c_3, c_4) = 0$.

80. Determine an algebraic condition on the coefficients of a quartic polynomial for that polynomial to have a quadruple root.

7. Quartic Equations in Geodesy

This section demonstrates how quartic equations arise in geodesy. The first application leads to a quartic equation for the *grazing angle,* also called the *angle of specular reflection,* between a ray incident from a source and its reflection off the surface of a sphere. To this end, consider a sphere S with equation $u^2 + v^2 + w^2 = 1$, and consider two points $\vec{\mathbf{P}}$ and $\vec{\mathbf{Q}}$ in space with coordinates (g, p, q) and (x, y, z), as in **Figure 5.**

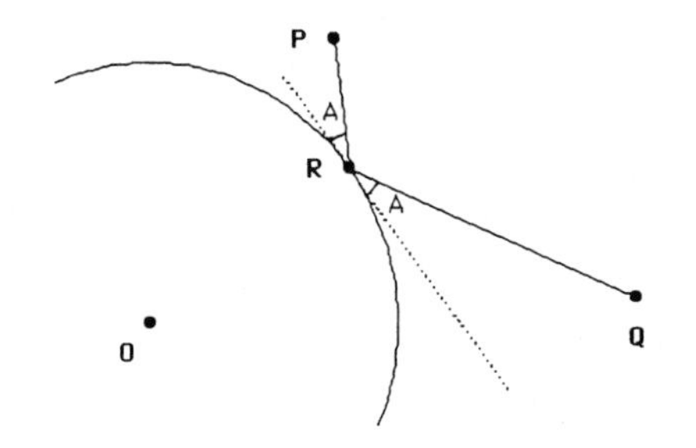

Figure 5. Grazing angle.

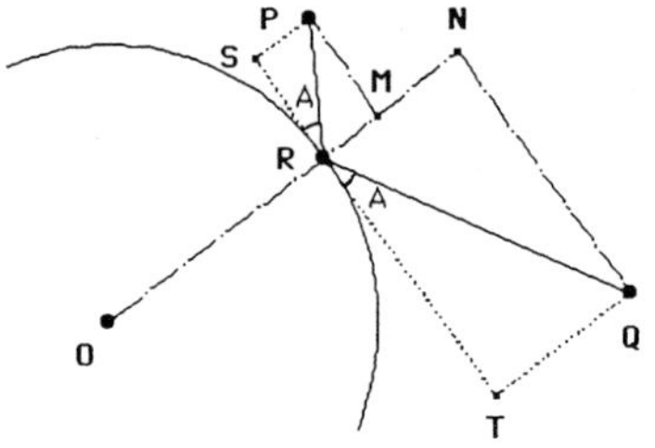

Figure 6. A determination of the grazing angle.

Also, consider a yet unknown point $\vec{\mathbf{R}}$ on the sphere S such that the two angles with the vertical, $\angle(\vec{\mathbf{P}}-\vec{\mathbf{R}},\vec{\mathbf{R}}-\vec{0})$ and $\angle(\vec{\mathbf{Q}}-\vec{\mathbf{R}},\vec{\mathbf{R}}-\vec{0})$, equal each other. The complementary angle A, with the tangent at $\vec{\mathbf{R}}$, is called the *grazing angle*. The problem consists in establishing an algorithm, formula, or method to compute $\vec{\mathbf{R}}$ and the grazing angle A in terms of the Cartesian coordinates (g,p,q) and (x,y,z).

The method of solution outlined here employs only plane geometry and dot products, without any trigonometry. To this end, observe that for the planar circular cross section (through $\vec{0}$, $\vec{\mathbf{P}}$, and $\vec{\mathbf{Q}}$) of a sphere with radius 1,

$\vec{\mathbf{R}}=(x,y)$ is also the unit normal to the sphere at $\vec{\mathbf{R}}$;

$i\vec{\mathbf{R}}=(-y,x)$ is the unit tangent to the sphere at $\vec{\mathbf{R}}$.

Moreover, consider the auxiliary points and lines introduced in **Figure 6**:

- $\vec{\mathbf{M}}$ is the orthogonal projection of $\vec{\mathbf{P}}$ on the normal to the sphere at $\vec{\mathbf{R}}$; thus,

$$\vec{\mathbf{M}}=\langle\vec{\mathbf{P}},\vec{\mathbf{R}}\rangle\vec{\mathbf{R}}.$$

- $\vec{\mathbf{N}}$ is the orthogonal projection of $\vec{\mathbf{Q}}$ on the normal to the sphere at $\vec{\mathbf{R}}$; thus,

$$\vec{\mathbf{N}}=\langle\vec{\mathbf{Q}},\vec{\mathbf{R}}\rangle\vec{\mathbf{R}}.$$

- $\vec{\mathbf{S}}$ is the orthogonal projection of $\vec{\mathbf{P}}$ on the tangent to the sphere at $\vec{\mathbf{R}}$; thus,

$$\vec{\mathbf{S}}=\langle\vec{\mathbf{P}},i\vec{\mathbf{R}}\rangle i\vec{\mathbf{R}}.$$

- $\vec{\mathbf{T}}$ is the orthogonal projection of $\vec{\mathbf{Q}}$ on the tangent to the sphere at $\vec{\mathbf{R}}$; thus,

$$\vec{\mathbf{T}}=\langle\vec{\mathbf{Q}},\vec{\mathbf{R}}\rangle i\vec{\mathbf{R}}.$$

By definition of the grazing angle, at the yet unknown point $\vec{\mathbf{R}}$ the triangles $\vec{\mathbf{P}}\vec{\mathbf{R}}\vec{\mathbf{S}}$ and $\vec{\mathbf{Q}}\vec{\mathbf{R}}\vec{\mathbf{T}}$ are congruent, whence the following two ratios are equal:

$$\frac{\overline{\vec{\mathbf{Q}}\vec{\mathbf{T}}}}{\overline{\vec{\mathbf{Q}}\vec{\mathbf{N}}}}=\frac{\overline{\vec{\mathbf{P}}\vec{\mathbf{S}}}}{\overline{\vec{\mathbf{P}}\vec{\mathbf{M}}}}.$$

Expressing the orthogonal projections in terms of dot products yields, with a change of sign because $\vec{\mathbf{P}}$ and $\vec{\mathbf{Q}}$ lie on opposite sides of the normal,

$$\frac{\langle \vec{\mathbf{Q}} - \vec{\mathbf{R}}, i\vec{\mathbf{R}} \rangle}{\langle \vec{\mathbf{Q}} - \vec{\mathbf{R}}, \vec{\mathbf{R}} \rangle} = -\frac{\langle \vec{\mathbf{P}} - \vec{\mathbf{R}}, i\vec{\mathbf{R}} \rangle}{\langle \vec{\mathbf{P}} - \vec{\mathbf{R}}, \vec{\mathbf{R}} \rangle},$$
$$\langle \vec{\mathbf{Q}} - \vec{\mathbf{R}}, i\vec{\mathbf{R}} \rangle \langle \vec{\mathbf{P}} - \vec{\mathbf{R}}, \vec{\mathbf{R}} \rangle = -\langle \vec{\mathbf{P}} - \vec{\mathbf{R}}, i\vec{\mathbf{R}} \rangle \langle \vec{\mathbf{Q}} - \vec{\mathbf{R}}, \vec{\mathbf{R}} \rangle.$$

With $\vec{\mathbf{R}} = (x, y)$, $\vec{\mathbf{P}} = (p_1, p_2)$, and $\vec{\mathbf{Q}} = (q_1, q_2)$, the latter equation becomes

$$(x^2 - y^2)(p_1 q_2 + p_2 q_1) + 2xy(p_2 q_2 - p_1 q_1) - x(q_2 + p_2) + y(p_1 + q_1) = 0.$$

Thus, $\vec{\mathbf{R}} = (x, y)$ lies at an intersection of the unit circle and a hyperbola passing through the origin. The substitution $x^2 + y^2 = 1$ and algebraic transformations yield

$$2x^2(p_1 q_2 + p_2 q_1) - x(p_2 + q_2) - (p_1 q_2 + p_2 q_1) = \sqrt{1 - x^2}\,[2x(p_2 q_2 - p_1 q_1) - (p_1 + q_1)],$$

whence squaring both sides produces a quartic equation for x with real coefficients.

The solution $\vec{\mathbf{R}} = (x, y)$ then represents the point where the surface reflects the ray from $\vec{\mathbf{Q}}$ toward $\vec{\mathbf{P}}$, or vice versa. Then

$$\cos(A) = \langle \vec{\mathbf{P}} - \vec{\mathbf{R}}, i\vec{\mathbf{R}} \rangle / |\vec{\mathbf{P}} - \vec{\mathbf{R}}|, \qquad \sin(A) = \langle \vec{\mathbf{P}} - \vec{\mathbf{R}}, \vec{\mathbf{R}} \rangle / |\vec{\mathbf{P}} - \vec{\mathbf{R}}|.$$

In contrast, an alternative method of solution produces a quartic equation for $\cos 2A$ with complex coefficients [Miller and Vegh 1993]. Yet either methods then requires an examination of the four solutions to select the one that corresponds to the point of reflection $\vec{\mathbf{R}}$ or the grazing angle A. Therefore, an iterative algorithm that provably converges to the desired solution would seem preferable to the quartic formulae—which also require iterative algorithms for the complex cube roots—but such a better algorithm does not seem to have appeared in print yet [Miller and Vegh 1993].

81. (†) This exercise leads to a quartic equation for the geodetic coordinates of a point in space, measured perpendicularly to the surface of an ellipsoid. To this end, consider an ellipsoid E with equation $(x/a)^2 + (y/a)^2 + (z/b)^2 = 1$, as in **Figure 7**, and consider a point P with Cartesian coordinates (u, v, w). The *geodetic altitude* of P is the Euclidean distance h from P to E, in other words, the minimum distance from P to any point on the ellipse. The *geodetic latitude* is the angle λ between the normal from the point P to the ellipse E and the equatorial plane (where $z = 0$). Establish an algorithm, formula, or method to compute the *geodetic coordinates* (h, λ) in terms of the Cartesian coordinates (u, v, w).

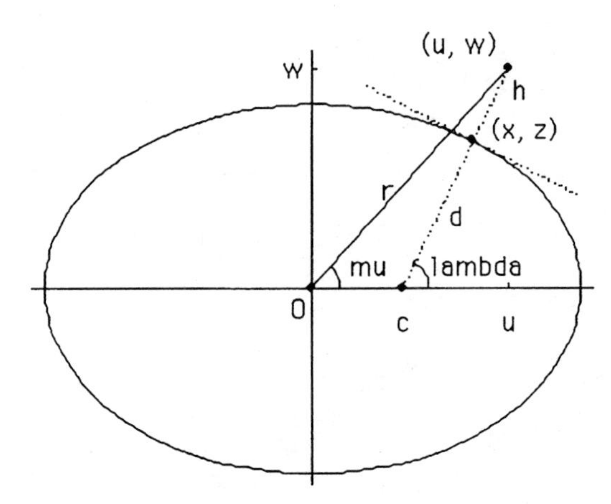

Figure 7. Geodetic coordinates.

82. (†) Establish an algorithm to calculate the grazing angle off an ellipse.

8. Solutions to Odd-Numbered Exercises

Exercise 1. $(1+2i)\cdot(3+4i) = 1\cdot 3 + 1\cdot 4i + 2i\cdot 3 + 2i\cdot 4i = 3 + 4i + 6i + 8i^2 = 3 + 10i - 8 = -5 + 10i$. ✓

Exercise 3. Denote the length by ℓ and the width by w, both expressed in yards. Then $2\ell + 2w = 10$, because the perimeter equals ten yards, whence $\ell + w = 5$. Moreover, $\ell w = 6$, because the area equals six square yards. The two equations just obtained form a system with the sum and the product of the two unknowns:

$$\begin{aligned} \ell + w &= 5, \\ \ell w &= 6. \end{aligned}$$

Several methods exist to solve such a system. For instance, solve the first equation for ℓ, which gives $\ell = 5 - w$. Then substitute the result into the second equation, which produces the quadratic equation $(5-w)w = 6$, whence a multiplication and a rearrangement give $w^2 - 5w + 6 = 0$. The quadratic formula then yields

$$w = \frac{5 \pm \sqrt{(-5)^2 - 4\cdot 1\cdot 6}}{2\cdot 1} = \frac{5 \pm \sqrt{25 - 24}}{2} = \frac{5 \pm 1}{2} \in \{2, 3\}.$$

Finally, return to the expression for $\ell = 5 - w$ to get $w = 2$ and $\ell = 5 - 2 = 3$, or $w = 3$ and $\ell = 5 - 3 = 2$. In both cases, the rectangular altar has one side equal to two yards and one side equal to three yards.

Exercise 5. Denote the height by h and the length by ℓ. Then the area of the façade becomes $h\ell = 1$, and the two ratios in question are $h/\ell = 2\ell/(2h+2\ell)$, whence the system

$$\begin{aligned} h\ell &= 1, \\ \frac{h}{\ell} &= \frac{\ell}{h+\ell}. \end{aligned}$$

Clearing the denominators in the second equation gives

$$h^2 + h\ell = \ell^2,$$

whence dividing both sides by ℓ^2 produces a quadratic equation for the ratio h/ℓ,

$$\left(\frac{h}{\ell}\right)^2 + \frac{h}{\ell} - 1 = 0.$$

Hence, solving for h/ℓ by means of the quadratic formula yields

$$\frac{h}{\ell} = \frac{-1 \pm \sqrt{1^2 + 41^2}}{2} = \frac{-1 \pm \sqrt{5}}{2}.$$

The positive solution and its reciprocal represent the classical *golden ratio*:

$$\frac{h}{\ell} = \frac{-1+\sqrt{5}}{2} = 0.618\,033\,988\ldots,$$
$$\frac{\ell}{h} = \frac{1+\sqrt{5}}{2} = 1.618\,033\,988\ldots.$$

Finally, from $h\ell = 1$ follows

$$h = \ell \cdot \frac{-1+\sqrt{5}}{2} = \frac{1}{h} \cdot \frac{-1+\sqrt{5}}{2},$$

and multiplying both sides by h yields

$$h = \sqrt{\frac{-1+\sqrt{5}}{2}} = 0.786\,151\,378\ldots,$$
$$\ell = \sqrt{\frac{1+\sqrt{5}}{2}} = 1.272\,019\,650\ldots.$$

Verify that $h\ell = 1$ and that $h/\ell = \ell/(h+\ell)$.

Exercise 7.

a) $(3,2)+(1,4) = (3+1, 2+4) = (4,6)$

b) $(1,4)+(3,2) = (1+3, 4+2) = (4,6)$

c) $(3,2)(1,4) = (3\cdot 1 - 2\cdot 4, 3\cdot 4 + 2\cdot 1) = (-5,14)$

d) $(1,4)(3,2) = (1\cdot 3 - 4\cdot 2, 1\cdot 2 + 4\cdot 3) = (-5,14)$

e) $(4,3)[(3,2)+(1,4)] = (4,3)(4,6) = (16-18, 24+12) = (-2,36)$

f) $[(4,3)(3,2)] + [(4,3)(1,4)] = (12-6, 8+9)+(4-12, 16+3) = (-2,36)$

g) $(4,3)^{-1} = (4/(4^2+3^2), -3/(4^2+3^2)) = ({}^4\!/\!{}_{25}, -{}^3\!/\!{}_{25})$

h) $[(3,2)+(1,4)]+(9,5) = (4,6)+(9,5) = (13,11)$

i) $(3,2)+[(1,4)+(9,5)] = (3,2)+(10,9) = (13,11)$

j) $[(3,2)(1,4)](9,5) = (-5,14)(9,5) = (-115,101)$

k) $(3,2)[(1,4)(9,5)] = (3,2)(-11,41) = (-115,101)$

l) $-(4,3) = (-4,-3)$

m) $(3,2)-(4,3) = (3,2)+(-4,-3) = (-1,-1)$

n) $(3,2)/(4,3) = (3,2)[(4,3)^{-1}] = (3,2)({}^4\!/\!{}_{25}, -{}^3\!/\!{}_{25}) = ({}^{18}\!/\!{}_{25}, -{}^1\!/\!{}_{25})$

Exercise 9. Associativity of complex addition: use the associativity of the real addition,

$$\begin{aligned}[(u,v)+(x,y)]+(p,q) &= (u+x,v+y)+(p,q) \quad = ([u+x]+p,[v+y]+q) \\ &= (u+[x+p],v+[y+q]) \quad = (u,v)+(x+p,y+q) \\ &= (u,v)+[(x,y)+(p,q)].\end{aligned}$$

Commutativity of complex addition: use the commutativity of the real addition,

$$(u,v)+(x,y)=(u+x,v+y)=(x+u,y+v)=(x,y)+(u,v).$$

Complex additive identity: use the real additive identity, 0,

$$(x,y)+(0,0)=(x+0,y+0)=(x,y)=(0+x,0+y)=(0,0)+(x,y).$$

Complex additive inverse: use the real additive inverse, 0,

$$(x,y)+(-x,-y)=(x-x,y-y)=(0,0)=(-x+x,-y+y)=(-x,-y)+(x,y).$$

Commutativity of complex multiplication: use the commutativity of the real multiplication,

$$(u,v)(x,y)=(ux-vy,uy+vx)=(xu-yv,yu+xv)=(x,y)(u,v).$$

Complex multiplicative identity: use the real multiplicative identity, 1,

$$(x,y)(1,0)=(x1-y0,x0+y1)=(x,y)=(1x-0y,0x+1y)=(1,0)(x,y).$$

Complex multiplicative inverse: use the real multiplicative inverse,

$$\begin{aligned}(x,y)\left(\frac{x}{x^2+y^2},\frac{-y}{x^2+y^2}\right) &= \left(\frac{x\cdot x-y\cdot(-y)}{x^2+y^2},\frac{x\cdot(-y)+yx}{x^2+y^2}\right)=(1,0) \\ &= \left(\frac{x\cdot x-(-y)\cdot y}{x^2+y^2},\frac{-y\cdot x+xy}{x^2+y^2}\right)\left(\frac{x}{x^2+y^2},\frac{-y}{x^2+y^2}\right)(x,y).\end{aligned}$$

Distributivity of complex multiplication over complex addition:

$$\begin{aligned}(u,v)[(x,y)+(p,q)] &= (u,v)(x+p,y+q) \\ &= (u[x+p]-v[y+q],u[y+q]+v[x+p]) \\ &= (ux+up-vy-vq,uy+uq+vx+vp), \\ [(x,y)+(p,q)](u,v) &= (x+p,y+q)(u,v) \\ &= ([x+p]u-[y+q]v,[y+q]u+[x+p]v) \\ &= (xu+pu-yv-qv,yu+qu+xv+pv) \\ &= (u,v)[(x,y)+(p,q)]. \quad \checkmark\end{aligned}$$

Exercise 11. $(-x,-y)+(x,y)=(-x+x,-y+y)=(0,0)$.

Exercise 13. The requirement that $(1,0) = (u,v)(x,y) = (ux - vy, uy + vx)$ means that $ux - vy = 1$ and $uy + vx = 0$. With the usual dot product in the plane, $\langle (h,\ell),(p,q)\rangle := hp + \ell q$, the second equation becomes

$$0 = uy + vx = \langle (u,v),(y,x)\rangle,$$

which means that (u,v) is perpendicular to (y,x), and, consequently, (u,v) is a real multiple of $(x,-y)$ of the form $(u,v) = r(x,-y) = (rx,-ry)$. Substituting this expression into the first equation gives $1 = ux - vy = (rx)x - (-ry)y = r(x^2+y^2)$. Consequently, $r = 1/(x^2+y^2)$, which yields the formula in **Table 1:** $(u,v) = (x/(x^2+y^2),\ -y/(x^2+y^2))$.

Exercise 15. See Higham [1996, 450].

Exercise 17. The existence of a multiplicative inverse with integral coordinates fails.

Exercise 19. Proceed by direct verifications, as for complex arithmetic. For the multiplicative inverse, verify that the modulus of the product equals the product of the moduli. See also Julstrom [1992].

Exercise 21.

a) $|(12,0)| = 12$

b) $|(0,5)| = 5$

c) $|(12,0) + (0,5)| = |(12,5)| = \sqrt{12^2 + 5^2} = \sqrt{169} = 13$

d) $|(12,0)| + |(0,5)| = 12 + 5 = 17$

e) $|(12,0)(0,5)| = 12 \cdot 5 = 60$

f) $|(12,0) - (0,5)| = |(12,-5)| = \sqrt{12^2 + (-5)^2} = \sqrt{169} = 13$

g) $|\,|(12,0)| - |(0,5)|\,| = |12 - 5| = 7$

h) $|(4,3)| = 5$

i) $|(12,5)| = \sqrt{12^2 + 5^2} = \sqrt{169} = 13$

j) $|(12,5)(4,3)| = 13 \cdot 5 = 65$

Exercise 23. Apply the triangle inequality repeatedly, as follows.

$$\begin{aligned} |w| &= |w + (-z + z)| = |(w - z) + z| \leq |w - z| + |z|, \\ |w| - |z| &\leq |w - z|; \\ |z| &= |z + (-w + w)| = |(z - w) + w| \leq |z - w| + |w|, \\ |z| - |w| &\leq |z - w|. \end{aligned}$$

However, $|z - w| = |-1|\,|z - w| = |(-1)(z - w)| = |w - z|$, and $|z| - |w| = -\left(|w| - |z|\right)$. Hence,

$$-\left(|w| - |z|\right) \leq |w - z| \leq |w| - |z|,$$

which means that

$$|w - z| \leq \left|\,|w| - |z|\,\right|.$$

Exercise 25.

$$\begin{aligned}
\overline{w+z} &= \overline{(u,v)+(x,y)} = \overline{(u+x,v+y)} = (u+x,-[v+y]) \\
&= (u,-v)+(x,-y)\overline{w}+\overline{z}, && \text{(C1)} \\
\overline{w\,z} &= \overline{(u,v)(x,y)} = \overline{(ux-vy,uy+vx)} = (ux-vy,-[uy+vx]) \\
&= (u,-v)(x,-y) = \overline{w}\,\overline{z}, && \text{(C2)} \\
z+\overline{z} &= (x,y)+(x,-y) = (x+x,0) = 2x = 2\mathrm{Re}(z), && \text{(C3)} \\
z-\overline{z} &= (x,y)-(x,-y) = (0,2y) = 2iy = 2i\mathrm{Im}(z), && \text{(C4)} \\
z\,\overline{z} &= (x,y)(x,-y) = (xx-y(-y),x(-y)+yx) = (x^2+y^2,0) = |z|^2. && \text{(C5)}
\end{aligned}$$

Exercise 27. Use the multiplicative property of the complex modulus. If $p = m^2 + n^2$ and $q = k^2 + \ell^2$, then $p = |(m,n)|^2$ and $q = |(k,\ell)|^2$, whence

$$\begin{aligned}
pq &= |(m,n)|^2 \cdot |(k,\ell)|^2 = |\{(m,n)\cdot(k,\ell)\}^2| = |(mk-n\ell,\ m\ell+nk)^2| \\
&= (mk-n\ell)^2 + (m\ell+nk)^2.
\end{aligned}$$

Exercise 29. With only real coefficients, it follows that $0 = p(z) = \overline{p(z)} = p(\overline{z})$:

$$\begin{aligned}
0 &= a_0 + a_1 z + a_2 z^2 + \cdots + a_{n-1} z^{n-1} + a_n z^n \\
&= \overline{a_0 + a_1 z + a_2 z^2 + \cdots + a_{n-1} z^{n-1} + a_n z^n} \\
&= a_0 + a_1 \overline{z} + a_2 \overline{z}^2 + \cdots + a_{n-1} \overline{z}^{n-1} + a_n \overline{z}^n.
\end{aligned}$$

Exercise 31. It suffices to verify that p has no root outside of the stated region. To this end, for every nonzero complex number $z \neq 0$, factoring out z and applying the inverse triangle inequality gives

$$\begin{aligned}
p(z) &= c_n z^n + \cdots + c_1 z + c_0 = z^n \left(c_n + \frac{c_{n-1}}{z} + \frac{c_{n-2}}{z^2} + \cdots + \frac{c_1}{z^{n-1}} + \frac{c_0}{z^n} \right), \\
|p(z)| &\geq |z|^n \left| |c^n| - \left| \frac{c_{n-1}}{z} + \frac{c_{n-2}}{z^2} + \cdots + \frac{c_1}{z^{n-1}} + \frac{c_0}{z^n} \right| \right|.
\end{aligned}$$

However,

$$\begin{aligned}
&\left| \frac{c_{n-1}}{z} + \frac{c_{n-2}}{z^2} + \cdots + \frac{c_1}{z^{n-1}} + \frac{c_0}{z^n} \right| \\
&\qquad \leq \max_{0 \leq k < n} \{|c_k|\} \left[\frac{1}{|z|} + \frac{1}{|z|^2} + \cdots + \frac{1}{|z|^n} \right] \\
&\qquad = \max_{0 \leq k < n} \{|c_k|\} \frac{1}{|z|} \frac{1 - (1/|z|)^n}{1 - (1/|z|)} \\
&\qquad \leq \max_{0 \leq k < n} \{|c_k|\} \frac{1}{|z| - 1}.
\end{aligned}$$

Consequently, if

$$|z| > 1 + \frac{1}{|c_n|} \max_{0 \le k < n} \{|c_k|\},$$

then

$$|c_n| > \max_{0 \le k < n} \{|c_k|\} \frac{1}{|z| - 1},$$

whence

$$\begin{aligned} |p(z)| &\ge |z|^n \left| |c^n| - \left| \frac{c_{n-1}}{z} + \frac{c_{n-2}}{z^2} + \cdots + \frac{c_1}{z^{n-1}} + \frac{c_0}{z^n} \right| \right| \\ &\ge |z|^n \left(|c^n| - \max_{0 \le k < n} \{|c_k|\} \frac{1}{|z| - 1} \right) > 0. \end{aligned}$$

For the inequality in the reverse direction, apply the foregoing result to $1/z$ and $c_n + c_{n-1}(1/z) + \cdots + c_1(1/z)^{n-1} + c_0(1/z)^n = 0$.

Exercise 33. First, $r = |(-\sqrt{3}, 1)| = \sqrt{(-\sqrt{3})^2 + 1^2} = \sqrt{4} = 2$. Second, $\cos\theta = x/r = -\sqrt{3}/2$ and $\sin\theta = y/r = 1/2$, whence $\theta = 5\pi/6$.

Exercise 35. From the definition of polar coordinates, $x = |z| \cos\theta = 5\cos(\pi/3) = 5\,(1/2) = 5/2$, and $y = |z| \sin\theta = 5\sin(\pi/3) = 5\sqrt{3}/2$; thus, $z = (5/2, 5\sqrt{3}/2)$.

Exercise 37. The complex cube roots of -8 are -2, $(1, -\sqrt{3})$, and $(1, \sqrt{3})$.

Exercise 39. All square roots w of $z \in \mathbb{C}$ satisfy the quadratic equation $w^2 - z = 0$. However, every quadratic polynomial has at most two affine factors, because the degree of their product, 2, is the sum of their degrees.

Exercise 41. Use direct calculations, multiplying out the dot product.

Exercise 43. Use direct calculations, expanding squared moduli in terms of dot products.

Exercise 45. Write $\gamma(t) = (X(t), Y(t))$, with $\gamma(1) = \gamma(0)$ for a closed curve, and apply Green's Theorem to each real coordinate:

$$\begin{aligned} \frac{1}{2i} \oint_{\partial\Omega} \bar{z}\, dz &= \frac{1}{2i} \int_0^1 \overline{\gamma(t)} \cdot \gamma'(t)\, dt \\ &= \frac{1}{2i} \int_0^1 (X(t), -Y(t)) \cdot (X'(t), Y'(t))(t)\, dt \\ &= \frac{1}{2i} \int_0^1 [(XX' + YY') + i(XY' - YX')](t)\, dt \\ &= \frac{1}{2i} \int_0^1 (X^2 + Y^2)'\, dt + \frac{1}{2} \int_0^1 (XY' - YX')(t)\, dt \\ &= \frac{1}{2i} \left(|\gamma(1)|^2 - |\gamma(0)|^2\right) + \frac{1}{2} \int\!\!\int_\Omega 2\, dx\, dy = 0 + A(\Omega). \end{aligned}$$

For an application to celestial mechanics, see Siegel and Moser [1995, 252].

Exercise 47. On a calculator working with ten decimal digits, the usual quadratic formula gives -0.0005 and 0. On a calculator working with twelve decimal digits, the usual quadratic formula gives -0.0005 and 2.5×10^{-15}. The calculators disagree on the second, nonnegative root; they do not produce any correct digit. In contrast, on the same calculators, the alternative quadratic formula, with **Algorithm 2**, gives the solutions to the proposed quadratic equation accurate to at least ten significant digits: -0.0005 and $1.999\,999\,999 \times 10^{-15}$. For the positive root X, the following calculation confirms that $1.999\,999\,999 \times 10^{-15} < X < 2 \times 10^{-15}$:

$$\begin{aligned}
2 \times 10^{-15} &> \frac{2 \cdot (2 \times 10^{-15})}{1 + \sqrt{1.000\,000\,000\,016}} = x_i = X \\
&> \frac{2 \cdot 2 \times 10^{-15}}{1 + \sqrt{1.000\,000\,000\,020\,000\,000\,000\,000\,1}} \\
&= \frac{2 \cdot 2 \times 10^{-15}}{1 + 1.000\,000\,000\,01} \\
&= \frac{2 \cdot 2 \times 10^{-15}}{2.000\,000\,000\,01} \\
&= \frac{2 \times 10^{-15}}{1.000\,000\,000\,005} \\
&= 2 \times 10^{-15} \left(1 - \{5 \times 10^{-12}\} + \{5 \times 10^{-12}\}^2 - \cdots\right) \\
&> 2 \times 10^{-15} \times (1 - 0.000\,000\,000\,005) \\
&= 1.999\,999\,999\,99 \times 10^{-15}
\end{aligned}$$

The last steps follow from expansion of the denominator, $1.000\,000\,000\,005$, according to the geometric series for $1/(1+h)$ with $h := 0.000\,000\,000\,005$.

Exercise 49.

a)

$$\begin{aligned}
u &= \sqrt{\frac{x + \sqrt{x^2 + y^2}}{2}} = \sqrt{\frac{5 + \sqrt{5^2 + 12^2}}{2}} = \sqrt{\frac{5 + 13}{2}} = \sqrt{\frac{18}{2}} \\
&= \sqrt{9} = 3, \\
v &= \frac{y}{2u} = \frac{12}{6} = 2, \\
(u, v) &= (3, 2); \\
\sqrt{(5, 12)} &= (3, 2).
\end{aligned}$$

As a verification, square the result just obtained:

$$(3, 2)^2 = (3, 2)(3, 2) = (9 - 4, 6 + 6) = (5, 12). \quad \checkmark$$

b)

$$\begin{aligned}
u &= \sqrt{\frac{x+\sqrt{x^2+y^2}}{2}} = \sqrt{\frac{7+\sqrt{7^2+24^2}}{2}} = \sqrt{\frac{7+25}{2}} = \sqrt{\frac{32}{2}} \\
&= \sqrt{16} = 4, \\
v &= \frac{y}{2u} = \frac{24}{8} = 3, \\
(u,v) &= (4,3); \\
\sqrt{(7,24)} &= (4,3).
\end{aligned}$$

As a verification, square the result just obtained:

$$(4,3)^2 = (4,3)(4,3) = (16-9, 12+12) = (7,24). \quad \checkmark$$

Exercise 51. No: try $(h,k) := (15,36)$, with $\ell = \sqrt{15^2+36^2} = 39$.

Exercise 53. Another complex square root exists, S, with $S(x,y) = (p,q)$, obtained by solving first for q instead of p, with

$$p = \text{sign}\,(y)\sqrt{\frac{x+\sqrt{x^2+y^2}}{2}}, \quad q = \sqrt{\frac{-x+\sqrt{x^2+y^2}}{2}}, \quad (p,q)^2 = (x,y).$$

Exercise 55. The equality $\sqrt{wz} = \sqrt{w}\sqrt{z}$ may fail, because different branches of the complex square root may appear on either side. For instance, with $w := -1 =: z$, the left-hand side becomes $\sqrt{(-1)(-1)} = \sqrt{1} = 1$, while the right-hand side becomes $\sqrt{-1}\sqrt{-1} = i \cdot i = -1$.

Exercise 57. Let $(u,v) := \sqrt{(x,y)}$, so that

$$\begin{aligned}
u &= \sqrt{\frac{\sqrt{x^2+y^2}+x}{2}}, \\
v &= \text{sign}(y)\sqrt{\frac{\sqrt{x^2+y^2}-x}{2}}, \\
u^2 - v^2 &= x, \\
u^2 + v^2 &= \sqrt{x^2+y^2}, \\
(2uv)^2 &= 4u^2v^2 = (u^2+v^2)^2 - (u^2-v^2)^2 = y^2.
\end{aligned}$$

For the principal branch of the complex square root, $u \geq 0$ and $\text{sign}(v) = \text{sign}(y)$, whence

$$2uv = y.$$

The general Cartesian equation of a straight line in the plane takes the form

$$px + qy = g$$

with $p^2 + q^2 = 1$, which corresponds to the line perpendicular to the direction (p, q) at the signed distance g from the origin. Consequently, if a point (x, y) lies on the line, then

$$\begin{aligned} px + qy &= g, \\ p(u^2 - v^2) + q(2uv) &= g, \\ p^2(u^2 - v^2) + 2pquv &= pg, \\ (pu + qv)^2 - (p^2 + q^2)v^2 &= pg, \\ (pu + qv)^2 - v^2 &= pg, \end{aligned}$$

which represents a hyperbola with center at the origin, one axis along the second (vertical) coordinate axis, and the other axis in the direction (p, q), perpendicular to the initial straight line.

a) For a horizontal line, $y = y_0$, whence $2uv = y_0$, which represents the half of a hyperbola in the first quadrant (for $y_0 > 0$) or in the second quadrant (for $y_0 < 0$).

b) For a vertical line in the right-hand half-plane, $x = x_0 > 0$, whence $u^2 - v^2 = x_0$, which lies on a rectangular hyperbola. For the principal branch of the complex square root, $u \geq 0$ and $\text{sign}(v) = \text{sign}(y)$, whence $\sqrt{\ }$ maps vertical lines from the right-hand half-plane into the half of the hyperbola in the same right-hand half-plane. If $x = x_0 < 0$, then $\sqrt{\ }$ splits vertical lines from the left-hand half-plane into two quarters of the hyperbola with $v^2 - u^2 = -x_0 > 0$ in the left-hand half-plane.

c) For a straight line through the origin, $px + qy = 0$, whence the general equation for the image becomes

$$(pu + qv)^2 - v^2 = 0, \qquad [(pu + qv) + v]\,[(pu + qv) - v]\,,$$

which corresponds to two straight lines through the origin. Thus, $\sqrt{\ }$ maps straight lines through the origin onto a broken line (one half of each of the two lines) in the right-hand half-plane, with vertex at the origin.

d) The general situation has already been handled above.

Exercise 59. For the particular circle with radius equal to zero, the principal branch of the complex square root $\sqrt{\ }$ maps the origin back to the origin. Hence the following considerations pertain to circles with positive radii. From $u^2 + v^2 = \sqrt{x^2 + y^2}$ it follows that $\sqrt{\ }$ maps the circle centered at the origin with radius r into the circle centered at the origin with radius $\sqrt{r}$. Moreover,

$$\begin{aligned} \cos[\text{Arg}\,(u, v)] &= \frac{u}{\sqrt{u^2 + v^2}} = \sqrt{\frac{1}{\sqrt{x^2 + y^2}}}\sqrt{\frac{\sqrt{x^2 + y^2} + x}{2}} \\ &= \sqrt{\frac{1 + x/\sqrt{x^2 + y^2}}{2}} = \cos[\text{Arg}\,(x, y)/2]. \end{aligned}$$

Therefore, the principal branch of the complex square root maps the arc with radius r between arguments $-\pi < \alpha \leq \beta \leq \pi$ to the arc with radius $\sqrt{r}$ between arguments $-\pi/2 < \alpha/2 \leq \beta/2 \leq \pi/2$. In contrast, $\sqrt{\ }$ maps arcs that cross the negative real axis, between arguments $\alpha < \pi < \beta$ with $\beta - \alpha < 2\pi$, to two arcs, one from $\alpha/2$ to $\pi/2$, and another from $-\pi/2$ to $-(\beta - \pi)/2$.

Exercise 61. To solve the quadratic equation $(1,3)z^2 + z - 1 = 0$, observe that $a = (1,3)$, $b = 1 = (1,0)$, and $c = -1 = (-1,0)$. Hence, compute

$$b^2 - 4ac = 1^2 - 4(1,3)(-1) = (1,0) + (4,12) = (5,12).$$

Also, from **Exercise 49a** follows $\sqrt{(5,12)} = (3,2)$. Moreover, $(1,3)^{-1} = (1/[1^2 + 3^2])(1,-3) = (1/10, -3/10)$. Consequently,

$$z = \frac{-b \pm \sqrt{b^2 - 4ac}}{2a} = \frac{-(1,0) \pm (3,2)}{2(1,3)}$$

whence arithmetic produces $z_1 = (2/5, -1/5)$ and $z_2 = (-1/2, 1/2)$.

Exercise 63. The solutions of the proposed cubic equation are -1, 2, and 3.

Exercise 65. The positive solution is approximately $(8.067 \times 10^{-6},\ 0)$. See also Macleod [1984] for an application to chemistry.

Exercise 67. Denote the height, length, and width by u, v, and w. Then the specifications translate into the following equations:

$$\begin{cases} u + v + w = 10, \\ uv + vw + uw = 31, \\ uvw = 30. \end{cases}$$

The equations just obtained also mean that u, v, and w are the three solutions of the cubic equation

$$z^3 - (u + v + w)z^2 + (uv + vw + wu)z - uvw = z^3 - 10z^2 + 31z - 30 = 0.$$

Cardano's formulae then yield the solutions: 2, 3, and 5, in any order, for instance,

$$u = 2, \quad v = 3, \quad w = 5.$$

Exercise 69. See Birkhoff and Mac Lane [1977, 120], Dickson [1914, 33–34], or Chandrasekharan [1985, 40–41].

Exercise 71. The solutions of the proposed cubic equation are -1, 2, and 3.

Exercise 73. See Grossman [1996, 143].

Exercise 75. From $z^4 - 2z^3 - 7z^2 + 8z + 12$ follows $t = -1/2$ and hence the reduced quartic $w^4 + (-17/2)w^3 + 0w + 225/16$. The resolvant cubic becomes $2d^3 + (-85/2)d^2 + (2087/8)d - 15827/32$, which has roots $d_0 := 49/4$, $d_1 := 17/4$, and $d_2 := 19/4$. To factor the initial quartic as a product of two real quadratics, only the root $d_0 = {}^{49}/_4$ satisfies the two inequalities $a + 2d \geq 0$ and $(a+d)^2 - c \geq 0$. With d_0, the reduced quartic factors as $(w^2 + 4w + {}^{15}/_4)(w^2 + 4w - {}^{15}/_4)$, whence with $z = w + {}^1/_2$ the initial quartic factors as

$$z^4 - 2z^3 - 7z^2 + 8z + 12 = (z^2 + 3z + 2)(z^2 - 5z + 6).$$

Exercise 77. See Meyer [1992] and Egerland and Hansen [1995].

Exercise 79. See Chandrasekharan [1985, 41–44].

Exercise 81. See Barrio and Riaguas [1993].

References

Ahlfors, L.V. 1979. *Complex Analysis.* 3rd ed. New York: McGraw-Hill.

Barrio, R., and A. Riaguas. 1993. Comparison of algorithms for the transformation from geocentric to geodetic coordinates. *Revista de la Academia de Ciencias Exactas, Físico-Químicas y Naturales de Zaragoza* Serie 2, 48: 135–143.

Birkhoff, G., and S. Mac Lane. 1977. *A Survey of Modern Algebra.* 4th ed. New York: Macmillan.

Boas, R.P. 1987. *Invitation to Complex Analysis.* New York: Random House.

Cannaday, E.R., P.F. Colwell, and H. Paley. 1986. Relevant and irrelevant internal rates of return. *The Engineering Economist* 32 (1): 17–38.

Cardano, G. 1993 [1545]. *Artis magnae, sive de regulis algebraicis. Lib. unus. Qui and totius operis de arithmetica, quod Opus Perfectum inscripsit, est in ordine decimus.* Translated by T. Richard Witmer as *Ars Magna, or the Rules of Algebra* 1993. Reprint. New York: Dover.

Chandrasekharan, K. 1985. *Elliptic Functions.* Berlin-Heidelberg-New York, NY: Springer-Verlag.

Churchill, Ruel V. and J.W. Brown. 1984. *Complex Variables and Applications.* 4th ed. New York: McGraw-Hill.

Chutsky, A. 1987. A more general definition of complex multiplication. *AMATYC Review* 8 (2): 19–22.

Dickson, L.E. 1914. *Elementary Theory of Equations.* London, UK: Wiley.

Egerland, W.O., and C.E. Hansen. 1995. Solution to Problem 10253. *American Mathematical Monthly* 102 (3): 277.

di Fagnano, G.C. 1750. *Produzioni mathematiche del Conte Gulio Carlo di Fagnano, Marchese de' Toschi, e di Sant' Onorio nobile romano, e patrizio senogagliese alla s antit a' din. s. Benedetto XIV. Pontefice Massimo.* Vol. 1. Pesaro, Italy: Gavelli.

Fenwick, E.H. 1992. Quaternions and the art of navigation. *International Journal of Mathematics Education in Science and Technology* 23 (2): 273–279.

Fisher, S.D. 1986. *Complex Variables.* Belmont and Monterey, CA: Wadsworth and Brooks/Cole.

Forsythe, George E. 1970. Pitfalls in computation, or why a math book isn't enough. *American Mathematical Monthly* 77 (9): 931–956.

Grossman, N. 1996. *The Sheer Joy of Celestial Mechanics.* Boston, MA: Birkhäuser.

Hahn, L. 1994. *Complex Numbers and Geometry.* Washington, DC: Mathematical Association of America.

Halberstam, H. and R.E. Ingram, eds. 1967. *The Mathematical Papers of Sir William Rowan Hamilton.* Vol. 3. Cambridge, UK: Cambridge University Press.

Hämmerlin, Günther, and K.-H. Hoffmann. 1991. *Numerical Mathematics.* New York, NY: Springer-Verlag.

Hamilton, W[illiam] R[owan]. 1834. On conjugate functions, or algebraic couples, as tending to illustrate generally the doctrine of imaginary quantities and as confirming the results of Mr Graves respecting the existence of two independent integers in the complete expression of an imaginary logarithm. *British Association Report*: 519–523.

Higham, N.J. 1996. *Accuracy and Stability of Numerical Algorithms.* Philadelphia, PA: Society for Industrial and Applied Mathematics (SIAM).

Hewlett-Packard. 1982. *HP-15C Owner's Handbook.* Part 00015–90001. Corvallis, OR: Hewlett-Packard Company.

________ . 1984. *HP-15C Advanced Functions Handbook.* Part 00015–90011, Rev. B. Corvallis, OR: Hewlett-Packard Company.

Hungerford, T.W. 1974. *Algebra.* New York: Holt, Rinehart and Winston.

Hut, P. 1985. Binary formation and interactions with field stars. In *Dynamics of Star Clusters,* edited by J. Goodman and P. Hut, 231–249. Princeton, NJ: International Astronomical Union.

Jacob, B. 1990. *Linear Algebra.* New York: Freeman.

Julstrom, B.A. 1992. *Using Real Quaternions to Represent Rotations in Three Dimensions.* UMAP Modules in Undergraduate Mathematics and Its Applications: Module 722. Reprinted in *The UMAP Journal* 13 (2) (1992): 121–148 and in *UMAP Modules: Tools for Teaching 1992,* edited by Paul J. Campbell, 1–34. Lexington, MA: COMAP, 1993. Comment on quaternions, by Donald Sullivan, *The UMAP Journal* 15 (1) (1994): 43–48.

Kahan, W.M. 1987. Branch cuts for complex elementary functions or much ado about nothing's sign bit. In *The State of the Art in Numerical Analysis,* edited by A. Iserles and M.J.D. Powell, 165–211. Oxford, UK: Clarendon Press.

Kelly, P. and Matthews, G. 1981. *The Non-Euclidean Hyperbolic Plane.* New York: Springer-Verlag.

Kincaid, D.R., and E.W. Cheney. 1996. *Numerical Analysis: The Mathematics of Scientific Computing.* 2nd ed. Pacific Grove, CA: Brooks/Cole.

Lang, S. 1965. *Algebra.* Reading, MA: Addison-Wesley.

Macleod, A.J. 1984. The calculation of pH. *International Journal of Mathematics Education in Science and Technology* 15: 691–696.

Meyer, W.W. 1992. Problem 10253. *American Mathematical Monthly* 99 (8): 782.

Miller, A.R., and E. Vegh. 1993. Exact result for the grazing angle of specular reflection from a sphere. *SIAM Review* 35: 472–480.

Muller, D.E. 1956. A method for solving algebraic equations using an automatic computer. *Mathematical Tables and Other Aids To Computation* 10: 208–215.

Narasimhan, R. 1985. *Complex Analysis in One Variable.* Boston, MA: Birkhäuser.

Nievergelt, Y. 1994. Exact equations for the equilibrium constants of single intermolecular complexes in terms of spectrophotometric data. *The Analyst* 119 (1): 145–151.

Paley, H., P.F. Colwell, and R.E. Cannaday. 1984. *Internal Rates of Return.* UMAP Modules in Undergraduate Mathematics and Its Applications: Module 640. Lexington, MA: COMAP. Reprinted in *UMAP Modules: Tools for Teaching 1983,* 493–548. Lexington, MA: COMAP, 1984.

Pence, D. 1984. *Spacecraft Attitude, Rotations and Quaternions.* UMAP Modules in Undergraduate Mathematics and Its Applications: Module 652. Reprinted in *The UMAP Journal* 5 (2): 215–250 and in *UMAP Modules: Tools for Teaching 1984,* edited by Paul J. Campbell, 129–172. Lexington, MA: COMAP, 1985. Lexington, MA: COMAP.

Pulskamp, Richard J., and James A. Delaney. 1991. *Computer and Calculator Computation of Elementary Functions.* UMAP Modules In Undergraduate Mathematics and Its Applications: Module 708. Lexington, MA: COMAP, 1991. Reprinted in *The UMAP Journal* 12 (1991): 315–348 and in *UMAP Modules: Tools for Teaching 1991,* edited by Paul J. Campbell, 1–34. Lexington, MA: COMAP, 1992.

Purcell, Edward M. 1963. *Electricity and Magnetism.* Berkeley Physics Course, Vol. 2. New York: McGraw-Hill.

Schletz, B. 1991. Use of quaternions in shuttle guidance, navigation, and control. *Mathematics Magazine* 64 (3): 172–175.

Schulz, W.C. 1982. Cubics with a rational root. *American Institute of Aeronautics and Astronautics.*

Schwerdtfeger, H. 1979. *Geometry of Complex Numbers: Circle Geometry, Moebius Transformations, Non-Euclidean Geometry.* New York: Dover.

Siegel, C. L., and J.K. Moser. 1995. *Lectures on Celestial Mechanics.* New York: Springer-Verlag.

Simmonds, J.G. 1996. Analytic functions, ideal fluid flow, and Bernoulli's equation. *SIAM Review* 38: 666–667.

Spiegel, M.R. 1964. *Complex Variables.* Schaum's Outline Series. New York: McGraw-Hill.

Stiefel, E.L., and G. Scheifele. 1971. *Linear and Regular Celestial Mechanics.* New York: Springer-Verlag.

Stoer, J., and R. Bulirsch. 1983. *Introduction to Numerical Analysis.* New York, NY: Springer-Verlag.

Struik, D.J. 1987. *A Concise History of Mathematics.* 4th revised ed. New York: Dover.

Sullivan, Donald. 1994. Comment on quaternions. *The UMAP Journal* 15 (1): 43–48.

van der Waerden, B.L. 1983. *Geometry and Algebra in Ancient Civilizations.* New York: Springer-Verlag.

________ . 1985. *A History of Algebra From al-Khwārizmi to Emmy Noether.* New York: Springer-Verlag.

About the Author

Yves Nievergelt graduated in mathematics from the École Polytechnique Fédérale de Lausanne (Switzerland) in 1976, with concentrations in functional and numerical analysis of PDEs. He obtained a Ph.D. from the University of Washington in 1984, with a dissertation in several complex variables under the guidance of James R. King. He now teaches complex and numerical analysis at Eastern Washington University.

Prof. Nievergelt is an associate editor of *The UMAP Journal.* He is the author of numerous UMAP Modules, a bibliography of case studies of applications of lower-division mathematics (*The UMAP Journal* 6 (2) (1985): 37–56), and *Mathematics in Business Administration* (Irwin, 1989).

UMAP

Modules in
Undergraduate
Mathematics
and Its
Applications

Published in
cooperation with

The Society for
Industrial and
Applied Mathematics,

The Mathematical
Association of America,

The National Council
of Teachers of
Mathematics,

The American
Mathematical
Association of
Two-Year Colleges,

The Institute for
Operations Research
and the Management
Sciences, and

The American
Statistical Association.

Module 756

Orthogonal Projections and Applications in Linear Algebra

Yves Nievergelt

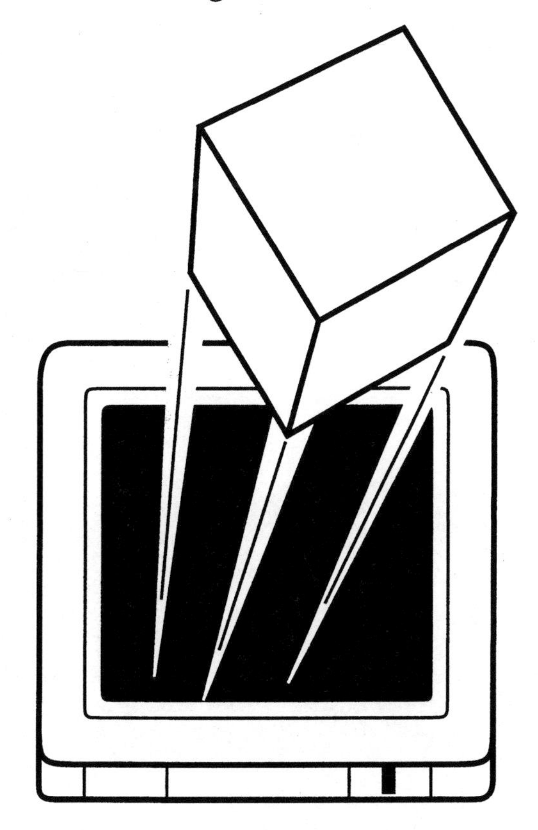

Applications of Computer Science, Engineering, Differential Equations, and Statistics to Linear Algebra

COMAP, Inc., Suite 210, 57 Bedford Street, Lexington, MA 02173 (781) 862–7878

INTERMODULAR DESCRIPTION SHEET: UMAP Unit 756

TITLE: Orthogonal Projections and Applications in Linear Algebra

AUTHOR: Yves Nievergelt
Department of Mathematics, MS 32
Eastern Washington University
526 5th Street
Cheney, WA 99004–2431
`ynievergelt@ewu.edu`

MATHEMATICAL FIELD: Linear algebra

APPLICATION FIELD: Computer science, engineering, differential equations, statistics

TARGET AUDIENCE: Students in linear algebra and subsequent courses.

ABSTRACT: This Module demonstrates how real applications require orthogonal projections in *abstract* linear spaces. The material presented here may serve either as a complement within a first course in linear algebra or as a review in a subsequent course, for instance, in computer graphics, numerical analysis, or Fourier analysis.

PREREQUISITES: A working knowledge of general inner products and Gram-Schmidt orthogonalization in general linear spaces. Some examples and exercises involve calculus.

RELATED UNITS:

Unit 324: *A Unified Method for Finding Laplace Transforms, Fourier Transforms, and Fourier Series*, by C. A. Grimm.

Unit 708: *Computer and Calculator Computation of Elementary Functions*, by Richard J. Pulskamp and James A. Delaney. Reprinted in *The UMAP Journal* 12 (1991): 315–348. Reprinted also in *UMAP Modules: Tools for Teaching 1991*, 1–34. Lexington, MA: COMAP, 1992.

Unit 717: *3-D Graphics in Calculus and Linear Algebra*, by Yves Nievergelt. Reprinted in *UMAP Modules: Tools for Teaching 1991*, 125–169. Lexington, MA: COMAP, 1992.

Unit 718: *Splines in Single and Multivariable Calculus*, by Yves Nievergelt. Reprinted in *UMAP Modules: Tools for Teaching 1992*, 39–101. Lexington, MA: COMAP, 1993.

COMAP, Inc., Suite 210, 57 Bedford Street, Lexington, MA 02173
(800) 77-COMAP = (800) 772-6627, or (781) 862-7878; `http://www.comap.com`

Orthogonal Projections and Applications in Linear Algebra

Yves Nievergelt
Department of Mathematics, MS 32
Eastern Washington University
526 5th Street
Cheney, WA 99004–2431
ynievergelt@ewu.edu

Table of Contents

Modules and Monographs in Undergraduate Mathematics and its Applications (UMAP) Project

The goal of UMAP is to develop, through a community of users and developers, a system of instructional modules in undergraduate mathematics and its applications, to be used to supplement existing courses and from which complete courses may eventually be built.

The Project was guided by a National Advisory Board of mathematicians, scientists, and educators. UMAP was funded by a grant from the National Science Foundation and now is supported by the Consortium for Mathematics and Its Applications (COMAP), Inc., a nonprofit corporation engaged in research and development in mathematics education.

1. Introduction

This Module demonstrates applications of orthogonal projections. The need for such a treatment arises in introductory linear algebra and in subsequent courses.

Many introductory linear algebra texts emphasize neither the ubiquity of orthogonal projections in subsequent courses, nor the importance of abstract orthogonal projections within general linear spaces in real applications.

Such texts may fail to provide sufficient motivation for the abstract theory required by some concrete applications. Also, subsequent courses may become more difficult than necessary when the need arises for orthogonal projections in general linear spaces over abstract fields, especially in applications in computer graphics, numerical analysis, or physics.

To alleviate these difficulties, this Module is a self-contained complement on orthogonal projections for linear algebra or subsequent courses.

In general, an "orthogonal projection" consists of a linear function $P : V \to W$ from a linear space V onto a linear subspace $W \subseteq P$, with the kernel (null space) of P perpendicular to the range W, and with P restricting to the identity function on W. One of the simplest applications of orthogonal projections lies in three-dimensional graphics (with or without computers), where V models the three-dimensional ambient space and W represents the computer screen. One of the many procedures to produce such graphics applies an orthogonal projection $P : V \to W$ to map objects in the space V onto the screen W, as demonstrated in the following section.

Nevertheless, many applications of orthogonal projections involve number fields different from the usual real numbers $\mathbf{R}$, and abstract linear spaces of functions instead of "vectors." Such applications include electronic signal analysis and synthesis, image compression, medical computed tomography, fast algebraic algorithms [Körner 1988; 1993], and partial differential equations in physics and engineering [Marti 1986].

Such applications depend upon the following concepts, all summarized near the end of this Module [Taylor 1958]:

- linear spaces,
- inner products,
- Cauchy's inequality,
- norms,
- triangle and reverse triangle inequalities,
- the Pythagorean theorem,
- the polar identity, and
- Bessel's inequality.

2. Applications of Orthogonal Projections

2.1 Application to Three-Dimensional Computer Graphics

For three-dimensional graphics, the ambient three-dimensional space may correspond to the linear space $V := \mathbf{R}^3$, in which a two-dimensional subspace $W \subset V$ represents the plane of the screen. To produce a picture of a point $\vec{\mathbf{x}} \in V$, a graphics procedure may then use an orthogonal projection $P : V \to W$ to map each point $\vec{\mathbf{x}}$ to its image $P(\vec{\mathbf{x}}) \in W$ on the screen.

Computationally, such a graphics procedure may endow the screen W with an orthonormal basis $(\vec{\mathbf{u}}, \vec{\mathbf{v}})$, then calculate the coordinates on the screen of the image $P(\vec{\mathbf{x}})$ by means of inner products,

$$P(\vec{\mathbf{x}}) = \langle \vec{\mathbf{x}}, \vec{\mathbf{u}} \rangle \vec{\mathbf{u}} + \langle \vec{\mathbf{x}}, \vec{\mathbf{v}} \rangle \vec{\mathbf{v}},$$

and finally draw the image of $\vec{\mathbf{x}}$ by plotting on the screen the point (p, q) with coordinates $p := \langle \vec{\mathbf{x}}, \vec{\mathbf{u}} \rangle$ and $q := \langle \vec{\mathbf{x}}, \vec{\mathbf{v}} \rangle$, as shown in **Figure 1**.

Example. The plane $W \subset \mathbf{R}^3$ that passes through the origin perpendicularly to the unit vector $\vec{\mathbf{w}} := (6/7, 2/7, -3/7)$ admits the orthonormal basis

$$\vec{\mathbf{u}} := (2/7, 3/7, 6/7), \quad \vec{\mathbf{v}} := (3/7, -6/7, 2/7).$$

Plotting the image of the point $\vec{\mathbf{x}} := (1, 2, 3)$ on the screen W then amounts to calculating the coordinates

$$p = \langle \vec{\mathbf{x}}, \vec{\mathbf{u}} \rangle = \langle (1,2,3), (2/7, 3/7, 6/7) \rangle = 26/7,$$
$$q = \langle \vec{\mathbf{x}}, \vec{\mathbf{v}} \rangle = \langle (1,2,3), (3/7, -6/7, 2/7) \rangle = -3/7.$$

Thus, the plotter would mark the point $(26/7, -3/7)$ to show on the screen the image of the point $(1, 2, 3)$ in space. □

For further details and additional examples, see, for instance, Nievergelt [1992a].

Remark 1. The foregoing outline in terms of inner products aims only at introducing the nature—as opposed to the details of computations—of three-dimensional computer graphics. Still, thanks to its simplicity, the approach just presented with inner products works well on very small machines, for instance, on pocket graphing calculators, as demonstrated by Nievergelt [1989]. In contrast, on larger machines the use of projective geometry also allows for perspectives and for rotations and translations with matrices, as explained by Hanes [1990]. □

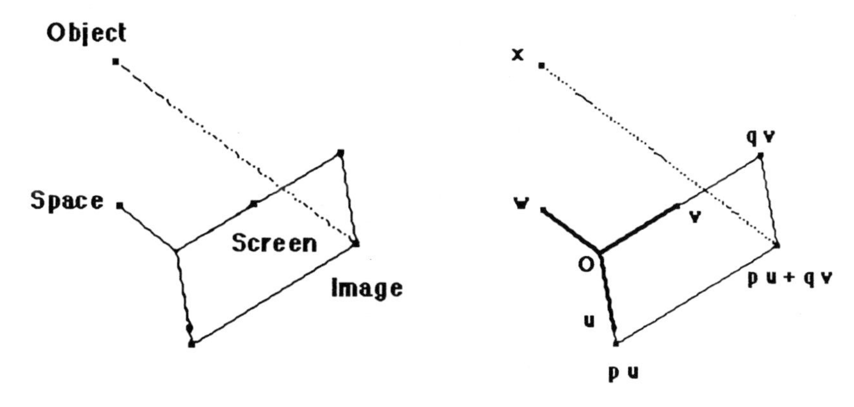

Figure 1. Three-dimensional graphics project objects in the three-dimensional ambient space onto a two-dimensional screen.

Exercises

1. In the three-dimensional space $V := \mathbf{R}^3$ with the usual inner product, the plane W passing through the origin perpendicularly to the unit vector $\vec{\mathbf{w}} := ({}^1\!/_{1469})(75, 180, 1456)$ admits the orthonormal basis

$$\vec{\mathbf{u}} := ({}^{12}\!/_{13}, -{}^5\!/_{13}, 0), \quad \vec{\mathbf{v}} := (-{}^{560}\!/_{1469}, -{}^{1344}\!/_{1469}, {}^{195}\!/_{1469}).$$

Calculate the orthogonal projection of the point $\vec{\mathbf{x}} := (1, 2, 3)$ on W.

2. In the same context as in the preceding exercise, calculate the orthogonal projection of the point $\vec{\mathbf{x}} = (1, 2, 3)$ on the plane spanned by as the orthonormal basis

$$\vec{\mathbf{u}} := (-{}^4\!/_5, {}^3\!/_5, 0), \quad \vec{\mathbf{v}} := (-{}^{12}\!/_{25}, -{}^{16}\!/_{25}, {}^{15}\!/_{25}).$$

2.2 Application to Ordinary Least-Squares Regression

The statistical method of ordinary least-squares (OLS) regression corresponds to an orthogonal projection of a vector of data on a linear subspace of specified vectors of coefficients.

Example. Consider the problem of fitting a straight line L with equation $c_1 x + c_0 = y$ to the data points

$$(2, 3), \quad (4, 7), \quad (5, 8), \quad (6, 9).$$

If all data points were on L, then they would satisfy the system

$$\begin{array}{rcrcl} 2c_1 & + & c_0 & = & 3, \\ 4c_1 & + & c_0 & = & 7, \\ 5c_1 & + & c_0 & = & 8, \\ 6c_1 & + & c_0 & = & 9, \end{array}$$

which, in terms of vectors, takes the equivalent form

$$c_1 \begin{pmatrix} 2 \\ 4 \\ 5 \\ 6 \end{pmatrix} + c_0 \begin{pmatrix} 1 \\ 1 \\ 1 \\ 1 \end{pmatrix} = \begin{pmatrix} 3 \\ 7 \\ 8 \\ 9 \end{pmatrix}.$$

Yet no solution exists, because the vector of data $\vec{\mathbf{y}} := (3, 7, 8, 9) \in \mathbf{Q}^4$ does not lie in the subspace $W := \text{Span}\{\vec{\mathbf{x}}, \vec{\mathbf{z}}\}$ spanned by the vectors $\vec{\mathbf{x}} := (2, 4, 5, 6)$ and $\vec{\mathbf{z}} := (1, 1, 1, 1)$. Therefore, the method of ordinary least-squares determines the linear combination $c_1\vec{\mathbf{x}} + c_0\vec{\mathbf{z}}$ closest to $\vec{\mathbf{y}}$ by means of the orthogonal projection $P : \mathbf{Q}^4 \to \text{Span}\{\vec{\mathbf{x}}, \vec{\mathbf{z}}\}$. For this example, ordinary least-squares regression yields the line displayed in **Figure 2**, from the following calculations.

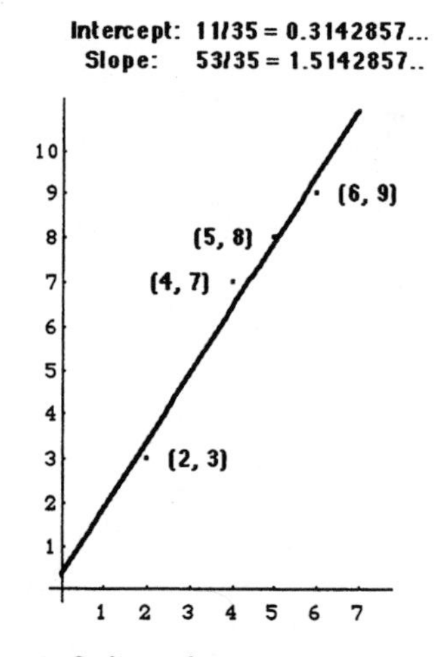

Figure 2. Ordinary least-squares regression.

Form the normal equations (reviewed in the summary and handout near the end of this module) $\langle \vec{\mathbf{w}}, \vec{\mathbf{x}}_i \rangle = \langle \vec{\mathbf{w}}, \vec{\mathbf{v}} \rangle$, which become

$$\begin{pmatrix} 2 & 4 & 5 & 6 \\ 1 & 1 & 1 & 1 \end{pmatrix} \begin{pmatrix} 2 & 1 \\ 4 & 1 \\ 5 & 1 \\ 6 & 1 \end{pmatrix} \begin{pmatrix} c_1 \\ c_0 \end{pmatrix} = \begin{pmatrix} 2 & 4 & 5 & 6 \\ 1 & 1 & 1 & 1 \end{pmatrix} \begin{pmatrix} 3 \\ 7 \\ 8 \\ 9 \end{pmatrix},$$

$$\begin{pmatrix} 81 & 17 \\ 17 & 4 \end{pmatrix} \begin{pmatrix} c_1 \\ c_0 \end{pmatrix} = \begin{pmatrix} 128 \\ 27 \end{pmatrix},$$

whence elimination, a calculator, or a computer yields

$$c_1 = \frac{53}{35} \approx 1.514\,285\,714\,285\dots,$$
$$c_0 = \frac{11}{35} \approx 0.314\,285\,714\,285\dots. \qquad \square$$

Remark 2. The foregoing example aims only at demonstrating that ordinary least-squares regression coincides with an orthogonal projection. However, the normal equations may exhibit an unacceptable sensitivity to the rounding errors that may occur in a digital computer. In principle, the algorithms described in the subsequent summary apply: Gram-Schmidt orthogonalization gives an orthogonal basis $(\vec{w}_1, \vec{w}_2)$ for $\text{Span}\{\vec{x}, \vec{z}\}$, whence inner products yield the projection $\langle\vec{y}, \vec{w}_1\rangle\vec{w}_1 + \langle\vec{y}, \vec{w}_2\rangle\vec{w}_2$, and then arithmetic converts the coefficients $\langle\vec{y}, \vec{w}_1\rangle$ and $\langle\vec{y}, \vec{w}_1\rangle$ back to c_1 and c_0. However, for ordinary least-squares, specialized methods yield greater speed and accuracy with digital computers, for instance, by means of matrix factorizations, as explained by Kincaid and Cheney [1996].

Exercises

3. Calculate the OLS line for the data points $(1,2), (2,6), (6,1)$.

4. Calculate the OLS line for the data points $(-7,1), (-1,5), (1,6), (7,8)$.

5. Calculate the OLS line for the data points $(1,3), (3,1), (4,5), (5,7), (7,4)$.

2.3 Application to the Computation of Functions

While digital computers can perform only finitely many arithmetic operations and logical tests, such transcendental functions as the exponential and trigonometric functions that occur in applications do not consist of finitely many such operations. Therefore, methods to approximate such transcendental functions with rational functions prove indispensable in scientific computing. One of the methods to approximate a transcendental function f with a rational function g involves a linear space V containing f and a subspace $W \subset V$ of rational functions of a specified degree n, with an inner product $\langle\ ,\ \rangle$. The method then determines the orthogonal projection g of f in W, so that g represents the rational function from W closest to f. For any prescribed accuracy, a degree n exists for which g approximates f to the specified accuracy, for instance, to all displayed digits on a calculator. In other words, g remains so close to f that the calculator computes g and displays the same result as it would have, had it computed f. Though a typical accuracy of 12 digits would require a large degree n, the following example illustrates the method just outlined for an accuracy of one significant digit.

Example. This example demonstrates how to determine a polynomial p of degree 1 in the design of a computer algorithm for the computation of the square root function. Consider the field $\mathbf{R}$ and the linear space $V := C^0([1/4, 1], \mathbf{R})$ of all real functions continuous on the interval $[1/4, 1]$, with the inner product

$$\langle f, g\rangle := \int_{1/4}^{1} f(x)g(x)\,dx.$$

For instance, the restriction of the square root function to that interval, $f : [1/4, 1] \to \mathbf{R}$, $f(x) := \sqrt{x}$, lies in the linear space $V = C^0([1/4, 1], \mathbf{R})$.

Also, consider the linear subspace $W \subset V$ consisting of all polynomials of degree at most 1 on the same interval, so that every $p \in W$ has the form $p(x) = c_0 + c_1 x$ for some coefficients $c_0, c_1 \in \mathbf{R}$. The problem examined here then amounts to calculating the orthogonal projection p of the square root f in the space W of affine polynomials. To this end, the first task lies in finding an orthonormal or orthogonal basis for W, for example, by applying the Gram-Schmidt Process to the basis $(v_1, v_2) := (1, x)$, which will utilize the following inner products:

$$\langle 1, 1\rangle = \int_{1/4}^{1} 1 \cdot 1\,dx = 3/4,$$

$$\langle x, 1\rangle = \int_{1/4}^{1} x \cdot 1\,dx = \left.\frac{x^2}{2}\right|_{1/4}^{1} = \frac{1^2 - (1/4)^2}{2} = \frac{16/16 - 1/16}{2} = \frac{15}{32},$$

$$\langle x, x\rangle = \int_{1/4}^{1} x \cdot x\,dx = \left.\frac{x^3}{3}\right|_{1/4}^{1} = \frac{1^3 - (1/4)^3}{3} = \frac{64/64 - 1/64}{3} = \frac{63}{252}.$$

Hence, Gram-Schmidt orthogonalization gives

$$w_1 = v_1 = 1,$$

$$w_2 = v_2 - \frac{\langle v_2, w_1\rangle}{\langle w_1, w_1\rangle} w_1 = x - \frac{\langle x, 1\rangle}{\langle 1, 1\rangle} 1 = x - \frac{15/32}{3/4} 1 = x - \frac{5}{8}.$$

Further, the optional orthonormalization yields

$$\langle w_2, w_2\rangle = \int_{1/4}^{1} (x - 5/8)^2\,dx = \left.\frac{(x - 5/8)^3}{3}\right|_{1/4}^{1} = \frac{9}{256} = \frac{3^2}{2^8},$$

$$u_1 = \frac{1}{\|w_1\|} w_1 = \frac{1}{\sqrt{3/4}} = \frac{2}{\sqrt{3}},$$

$$u_2 = \frac{1}{\|w_2\|} w_2 = \frac{\sqrt{256}}{\sqrt{9}} (x - 5/8) = \frac{16}{3} (x - 5/8).$$

In general, the use of the orthogonal basis (w_1, w_2) avoids the square roots involved in the orthonormal basis (u_1, u_2). Thus, the calculation of the orthogonal projection of f on W proceeds as follows.

$$\begin{aligned}
\langle f, w_1 \rangle &= \int_{1/4}^{1} \sqrt{x} \cdot 1\, dx = \left.\frac{x^{3/2}}{3/2}\right|_{1/4}^{1} = \frac{1 - 1/8}{3/2} = \frac{7}{12} = \frac{7}{2^2 \times 3}, \\
\langle f, v_2 \rangle &= \int_{1/4}^{1} \sqrt{x} \cdot x\, dx = \left.\frac{x^{5/2}}{5/2}\right|_{1/4}^{1} = \frac{1 - 1/32}{5/2} = \frac{31}{80} = \frac{31}{2^4 \times 5}, \\
\langle f, w_2 \rangle &= \langle f, v_2 - 5/8 w_1 \rangle \\
&= \langle f, v_2 \rangle - (5/8)\langle f, w_1 \rangle \\
&= 31/80 - (5/8)\,7/12 = \frac{11}{480} = \frac{11}{2^5 \times 3 \times 5}.
\end{aligned}$$

Hence, the affine polynomial p closest to the square root f on the interval $[1/4, 1]$ takes the form

$$\begin{aligned}
p &= \frac{\langle f, w_1 \rangle}{\langle w_1, w_1 \rangle} w_1 + \frac{\langle f, w_2 \rangle}{\langle w_2, w_2 \rangle} w_2 \\
&= \frac{7/12}{3/4} 1 + \frac{11/480}{9/256} (x - 5/8) \\
&= \frac{11 \times 8}{5 \times 27} x + \frac{7}{9} - \frac{11/480}{9/256}\frac{5}{8} \\
&= \frac{88}{135} x + \frac{10}{27} \\
&\approx (0.651\,851\,851\,851\ldots)x + 0.370\,370\,370\,370\ldots.
\end{aligned}$$

Thus,

$$\begin{aligned}
c_0 &= 10/27 \approx 0.370\,370\,370\,370\ldots, \\
c_1 &= 88/135 \approx 0.651\,851\,851\,851\ldots.
\end{aligned}$$

The result appears in **Figure 3.**

An analysis of the discrepancy $D(x) := c_0 + c_1 x - \sqrt{x}$ may provide an estimate of the accuracy achieved by the approximation just obtained. On the open interval $]1/4, 1[$, calculus shows that the discrepancy D has a local minimum at $x_* := 1/(4 \cdot c_1^2) \approx 0.588\,358\,729\,339\ldots$, where $D(x_*) = c_0 - [1/(4c_1)] = -125/9504 \approx -0.013\,152\,357\ldots$, $D'(x_*) = 0$ and $D''(x_*) > 0$. However, at the endpoints, $D(1) = 1/45 = 0.022\,222\ldots$ and $D(1/4) = 1/30 = 0.033\,333\ldots$. Consequently, the absolute discrepancy reaches its maximum at $1/4$. Because the square root also has its minimum absolute value there, the relative discrepancy attains its maximum at the same endpoint. Thus,

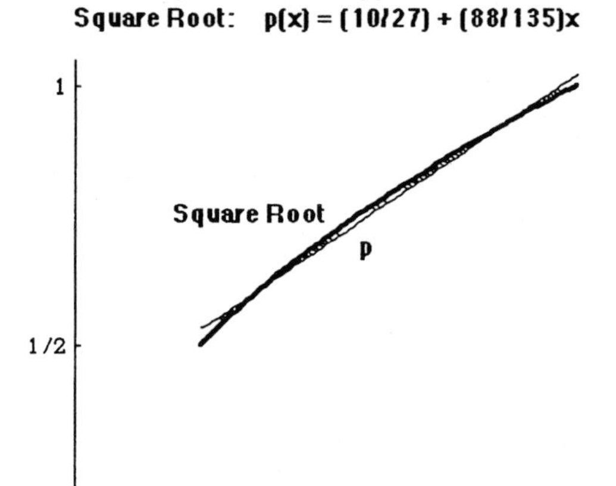

Figure 3. A least-squares affine approximation of the square root.

$$\frac{|c_0 + c_1 x - \sqrt{x}|}{|\sqrt{x}|} \leq \frac{|c_0 + c_1 (1/4) - \sqrt{1/4}|}{|\sqrt{1/4}|} \leq \frac{1/30}{1/2} = \frac{1}{15} = 0.066\,666\ldots .$$

□

The affine polynomial $p(x) = c_0 + c_1 x$ just obtained may be used to approximate the square root $\sqrt{\ }$ on the interval $[1/4, 1]$.

Example. If $x := 9/25$ then $\sqrt{9/25} = 3/5 = 0.6$ and $p(9/25) = c_0 + c_1 9/25 \approx 0.605\,037\,037\,037\ldots \approx 0.6$. □

The affine polynomial $p(x) = c_0 + c_1 x$ may also be used to approximate the square root $\sqrt{\ }$ on all of $\mathbf{R}$.

Example. If $z := 9$ then $\sqrt{9} = 3$. Moreover, $z = 9 = 25 \times 9/25$ with $9/25 \in [1/4, 1]$. Hence $p(9/25) = c_0 + c_1 9/25 \approx 0.605\,037\,037\,037\ldots$ and

$$\begin{aligned}\sqrt{9} = \sqrt{25 \times 9/25} = \sqrt{25} \times \sqrt{9/25} &\approx \sqrt{25} \cdot p(9/25) \\ &\approx 5 \times 0.605\,037\,037\,037\ldots \\ &\approx 3.025\,185\,185\ldots .\end{aligned}$$

Thus, a one-digit mini-calculator could evaluate the polynomial p, which requires only one multiplication and one addition, to compute $\sqrt{9}$ to all displayed digits. □

Remark 3. Many other methods exist to approximate functions to any degree of accuracy, for example, splines [Kincaid and Cheney 1996; Nievergelt 1992b] and Chebyshev's least absolute value approximation [Kincaid and Cheney 1996]. Algorithms for computers may follow such approximation by Newton's method to produce greater accuracy [Pulskamp and Delaney 1992]. □

Exercises

6. Consider the field $\mathbf{R}$ and the linear space $V := C^0([1/8, 1], \mathbf{R})$ consisting of all real functions continuous on the interval $[1/8, 1]$, with the inner product

$$\langle f, g \rangle := \int_{1/8}^{1} f(x) g(x)\, dx.$$

a) Apply Gram-Schmidt orthonormalization to the subset $\{1, x\}$.

b) In the subspace $W := \text{Span}\{1, x\}$, determine the affine function q of the form $q(x) = c_0 + c_1 x$ closest to the cube root function $\sqrt[3]{\ } \in V$.

c) For a verification, compare $\sqrt[3]{64/125}$ with its approximation $q(64/125)$.

d) Explain how to use the polynomial q to approximate the cube root of each real number, and verify your procedure with $\sqrt[3]{27}$.

e) Estimate the maximum relative discrepancy between the cube root and its approximation q on $[1/8, 1]$.

7. This problem demonstrates a method to design an algorithm to compute the exponential function. To this end, consider the linear space $V := C^0([-1, 0], \mathbf{R})$ of all continuous real-valued functions defined on the closed interval $[-1, 0]$, with the inner product

$$\langle f, g \rangle := \int_{-1}^{0} f(x) \cdot g(x)\, dx.$$

Also, consider the function $\exp_2 : [-1, 0] \to \mathbf{R}$ defined by $\exp_2(x) := 2^x$.

a) Calculate the orthogonal projection, denoted by g, of the function $\exp_2$ just defined on the linear subspace $W := \text{Span}\{1, x\}$.

b) For a verification, compare $\exp_2(-1/2)$ with its approximation $g(-1/2)$.

c) Explain how to utilize g to approximate e^x for each $x \in \mathbf{R}$, and test your procedure with e^1.

d) Estimate the maximum relative discrepancy between $\exp_2$ and its approximation g.

2.4 Application to Fourier Series and Transforms

Fourier series and Fourier transforms constitute orthogonal projections of functions on subspaces spanned by orthonormal bases of exponential or trigonometric functions.

Definition 1. A function $f : \mathbf{R} \to \mathbf{C}$ has **period** $T > 0$ if, but only if, $f(t+T) = f(t)$ for each $t \in \mathbf{R}$. □

Example: Continuous Real Fourier Series. Consider the real linear space $V := C^0_T(\mathbf{R}, \mathbf{R})$ of all continuous functions $f : \mathbf{R} \to \mathbf{R}$ with period $T > 0$, endowed with the inner product

$$\langle f, g \rangle := \frac{2}{T} \int_A^{A+T} f(t) \cdot g(t)\, dt,$$

which gives the same result regardless of $A \in \mathbf{R}$ by periodicity. In the linear space $C^0_T(\mathbf{R}, \mathbf{R})$, straightforward calculus shows that the following functions form an orthonormal set:

$$\frac{1}{2},\ \cos\left(\frac{2\pi}{T} \cdot t\right),\ \sin\left(\frac{2\pi}{T} \cdot t\right),\ \cos\left(2 \cdot \frac{2\pi}{T} \cdot t\right),\ \sin\left(2 \cdot \frac{2\pi}{T} \cdot t\right),$$
$$\ldots,\ \cos\left(k \cdot \frac{2\pi}{T} \cdot t\right),\ \sin\left(k \cdot \frac{2\pi}{T} \cdot t\right),\ \ldots.$$

Consequently, the linear combination of the first $2N+1$ such functions that approximates a function f most closely, in the sense of least squares, is the orthogonal projection of f onto the linear subspace spanned by these functions,

$$f \approx a_0 \cdot \frac{1}{2} + \sum_{k=1}^{N} \left[a_k \cdot \cos\left(k \cdot \frac{2\pi}{T} \cdot t\right) + b_k \cdot \sin\left(k \cdot \frac{2\pi}{T} \cdot t\right) \right],$$

with coefficients

$$a_k := \left\langle f, \cos\left(k \cdot \frac{2\pi}{T} \cdot t\right) \right\rangle = \frac{2}{T} \int_A^{A+T} f(t) \cdot \cos\left(k \cdot \frac{2\pi}{T} \cdot t\right) dt,$$
$$b_k := \left\langle f, \sin\left(k \cdot \frac{2\pi}{T} \cdot t\right) \right\rangle = \frac{2}{T} \int_A^{A+T} f(t) \cdot \sin\left(k \cdot \frac{2\pi}{T} \cdot t\right) dt. \quad \square$$

Example. Consider the "characteristic" function χ of the closed interval $[-\pi/2, \pi/2]$, displayed at the top of **Figure 4** and defined by

$$\chi(t) := \begin{cases} 1, & \text{if } t \in [-\pi/2, \pi/2]; \\ 0, & \text{if } t \notin [-\pi/2, \pi/2]. \end{cases}$$

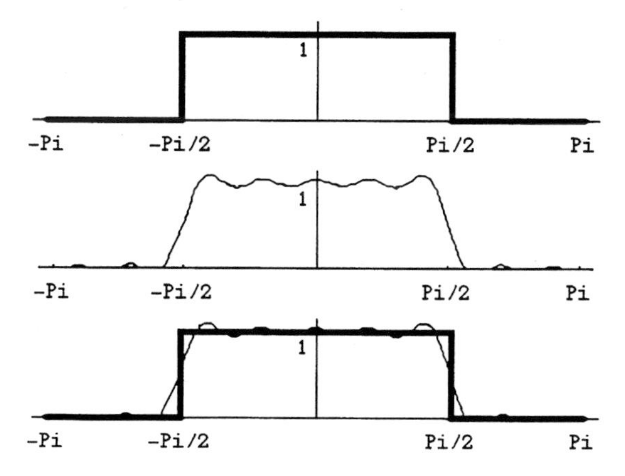

Figure 4. A least-squares approximation of a step function by the first ten terms of its Fourier series.

Taking advantage of "odd" and "even" integrands, calculus gives

$$\begin{aligned}
a_0 &= \frac{1}{\pi}\int_{-\pi}^{\pi} \chi(t)\cdot 1\,dt = \frac{1}{\pi}\int_{-\pi/2}^{\pi/2} 1\cdot 1\,dt = \frac{2}{\pi}\int_{0}^{\pi/2} 1\cdot 1\,dt = 1,\\
a_k &= \frac{1}{\pi}\int_{-\pi}^{\pi} \chi(t)\cdot \cos(k\cdot t)\,dt = \frac{2}{\pi}\int_{0}^{\pi/2} 1\cdot \cos(k\cdot t)\,dt\\
&= \frac{2}{\pi}\cdot \left.\frac{\sin(k\cdot t)}{k}\right|_0^{\pi/2} = \frac{2}{\pi}\cdot\frac{\sin(k\cdot \pi/2)}{k},\\
b_k &= \frac{1}{\pi}\int_{-\pi}^{\pi} \chi(t)\cdot \sin(k\cdot t)\,dt = 0.
\end{aligned}$$

Thus, the linear combination of such trigonometric functions closest to the characteristic function becomes

$$\chi(t) \approx \frac{1}{2} + \frac{2}{\pi}\cdot\left[\frac{1}{1}\cdot\cos(t) + \frac{-1}{3}\cdot\cos(3t) + \frac{1}{5}\cdot\cos(5t) + \frac{-1}{7}\cdot\cos(7t) + \frac{1}{9}\cdot\cos(9t) + \cdots\right].$$

The sum of the first ten terms appears in **Figure 4.** □

Example: Continuous Complex Fourier Series. Consider the complex linear space $V := C_T^0(\mathbf{R}, \mathbf{C})$ of all continuous functions $f : \mathbf{R} \to \mathbf{C}$ with period $T > 0$, endowed with the inner product

$$\langle f, g\rangle := \frac{1}{T}\int_A^{A+T} f(t)\cdot\overline{g(t)}\,dt,$$

which gives the same result regardless of $A \in \mathbf{R}$ by periodicity. In the linear space $C_T^0(\mathbf{R}, \mathbf{C})$, straightforward calculus shows that the following functions form an orthonormal set:

$$\left\{t \mapsto \exp\left(i \cdot k \cdot \frac{2\pi}{T} \cdot t\right) : k \in \mathbf{Z}\right\} =$$

$$\left\{\ldots, \exp\left(-i \cdot k \cdot \frac{2\pi}{T} \cdot t\right), \ldots, \exp\left(-i \cdot 2 \cdot \frac{2\pi}{T} \cdot t\right), \quad \exp\left(-i \cdot \frac{2\pi}{T} \cdot t\right),\right.$$

$$\left.1, \quad \exp\left(i \cdot \frac{2\pi}{T} \cdot t\right), \quad \exp\left(i \cdot 2 \cdot \frac{2\pi}{T} \cdot t\right), \ldots, \exp\left(i \cdot k \cdot \frac{2\pi}{T} \cdot t\right), \ldots\right\}.$$

Consequently, the linear combination of the first $2N + 1$ such functions that approximates a function f most closely, in the sense of least squares, is the orthogonal projection of f onto the linear subspace spanned by these functions,

$$f \approx \sum_{k=-N}^{N} c_k \cdot \exp\left(i \cdot k \cdot \frac{2\pi}{T} \cdot t\right),$$

with coefficients

$$c_k := \left\langle f, \exp\left(i \cdot k \cdot \frac{2\pi}{T} \cdot t\right)\right\rangle = \frac{1}{T} \int_A^{A+T} f(t) \cdot \exp\left(-i \cdot k \cdot \frac{2\pi}{T} \cdot t\right) dt.$$

□

Example: Discrete Complex Fourier Transform. Consider the complex linear space $V := \mathbf{C}^{\mathbf{N}}$ of all functions $f : \{0, 1, \ldots, N-2, N-1\} \to \mathbf{C}$ with the inner product

$$\langle f, g \rangle_N := \frac{1}{N} \sum_{\ell=0}^{N-1} f(\ell) \cdot \overline{g(\ell)}.$$

In the linear space $\mathbf{C}^{\mathbf{N}}$, straightforward calculations with geometric series show that the following functions form an orthonormal set:

$$\{\ell \mapsto \exp(i \cdot k \cdot \ell) : k \in \{0, 1, \ldots, N-2, N-1\}\} =$$

$$\{1, \exp(i \cdot \ell), \exp(i \cdot 2 \cdot \ell), \ldots, \exp(i \cdot [N-2] \cdot \ell), \exp(i \cdot [N-1] \cdot \ell)\}.$$

Consequently, the linear combination of these functions that approximates a function f most closely, in the sense of least squares, is the orthogonal projection of f onto the linear subspace spanned by these functions,

$$f \approx \sum_{k=0}^{N-1} c_k \cdot \exp(i \cdot k \cdot \ell),$$

with coefficients

$$c_k := \langle f, \exp(i \cdot k \cdot \ell) \rangle_N = \frac{1}{N} \sum_{\ell=0}^{N-1} f(\ell) \cdot \exp(-i \cdot k \cdot \ell).$$

□

Exercises

The following exercises demonstrate by examples how the accuracy of the partial sums of the Fourier series increases as the degree of differentiability of the function increases: *the smoother the function, the more accurate its approximations by partial sums of its Fourier series.*

8. Calculate the Fourier coefficients of the function $f : \mathbf{R} \to \mathbf{R}$ defined by

$$f(x) := \begin{cases} 1, & \text{if } -\pi \le x < -\pi/2; \\ -1, & \text{if } -\pi/2 \le x < \pi/2; \\ 1, & \text{if } \pi/2 \le x < \pi; \end{cases}$$

and extended to have period 2π.

9. Calculate the Fourier coefficients of the function $g : \mathbf{R} \to \mathbf{R}$ defined by

$$g(x) := \begin{cases} x + \pi, & \text{if } -\pi \le x < -\pi/2; \\ -x, & \text{if } -\pi/2 \le x < \pi/2; \\ x - \pi, & \text{if } \pi/2 \le x < \pi; \end{cases}$$

and extended to have period 2π.

10. Calculate the Fourier coefficients of the function $h : \mathbf{R} \to \mathbf{R}$ defined by

$$h(x) := \begin{cases} (1/2)(x + \pi)^2, & \text{if } -\pi \le x < -\pi/2; \\ (\pi^2/4) - (x^2/2), & \text{if } -\pi/2 \le x < \pi/2; \\ (1/2)(x - \pi)^2, & \text{if } \pi/2 \le x < \pi; \end{cases}$$

and extended to have period 2π.

A theorem from Fourier analysis states that if a function F is continuous at a point x, then its Fourier series converges to $F(x)$ at x. The following exercises demonstrate the use of such a theorem in the calculation of some infinite series.

11. With f as in **Exercise 8**, verify that if the Fourier series of f converges to $f(0)$ at 0, then

$$\pi = 4\left[1 - \frac{1}{3} + \frac{1}{5} - \frac{1}{7} + \frac{1}{9} + \cdots\right] = 4\sum_{m=1}^{\infty} \frac{(-1)^m}{2m+1}.$$

12. With h as in **Exercise 10**, verify that if the Fourier series of h converges to $h(0)$ at 0, then

$$\pi^3 = 32\left[1 - \frac{1}{27} + \frac{1}{125} - \frac{1}{343} + \frac{1}{729} + \cdots\right] = 32\sum_{m=1}^{\infty} \frac{(-1)^m}{(2m+1)^3}.$$

3. Summary of Definitions and Theorems

3.1 Number Fields

Definition 2. A **number field** consists of a set $\mathbf{F}$ containing at least two distinct elements $0_{\mathbf{F}}$ and $1_{\mathbf{F}}$, with two binary operations $+$ and $\times$, which are functions

$$\begin{aligned} + &: \mathbf{F} \times \mathbf{F} \to \mathbf{F}, \quad (r,s) \mapsto r+s, \\ \times &: \mathbf{F} \times \mathbf{F} \to \mathbf{F}, \quad (r,s) \mapsto r \times s \quad (\text{or} \quad r \cdot s \quad \text{or} \quad rs), \end{aligned}$$

such that all the properties in **Table 1** hold. □

Table 1. Algebraic properties of number fields.

The following properties must hold for all elements $h, k, \ell \in \mathbf{F}$.

Property	
(1) Associativity of $+$	$[h+\ell]+k = h+[\ell+k]$
(2) Commutativity of $+$	$h+\ell = \ell+h$
(3) Additive identity	$\ell + 0_{\mathbf{F}} = \ell = 0_{\mathbf{F}} + \ell$
(4) Additive inverse	$\mathbf{F}$ contains k with $\ell + k = 0_{\mathbf{F}} = k + \ell$
(5) Associativity of $\times$	$[h\ell]k = h[\ell k]$
(6) Commutativity of $\times$	$h\ell = \ell h$
(7) Multiplicative identity	$\ell 1_{\mathbf{F}} = \ell = 1_{\mathbf{F}}\ell$
(8) Multiplicative inverse	If $\ell \neq 0_{\mathbf{F}}$, then $\mathbf{F}$ contains k with $k\ell = 1_{\mathbf{F}}$
(9) Distributivity	$h[\ell+k] = [h\ell]+[hk]$

3.2 Linear Spaces

Definition 3. A **linear space over a field $\mathbf{F}$** consists of a set V containing at least one element 0_V, with two binary operations $+$ and $\cdot$, which are functions

$$\begin{aligned} + &: V \times V \to V, \quad (v,w) \mapsto v+w, \\ \cdot &: \mathbf{F} \times V \to V, \quad (r,v) \mapsto r \cdot v \quad (\text{or} \quad rv), \end{aligned}$$

such that all the properties in **Table 2** hold. □

Table 2. Algebraic properties of linear spaces.

The following properties must hold for all elements $r, s \in \mathbf{F}$ and $u, v, w \in V$.

(1) Associativity of $+$	$[u + v] + w = u + [v + w]$
(2) Commutativity of $+$	$u + vl = v + u$
(3) Additive identity	$v + 0_V = v = 0_V + v$
(4) Additive inverse	$\mathbf{F}$ contains w with $v + w = 0_V = w + v$
(5) Associativity of $\cdot$	$[rs] \cdot k = r \cdot [s \cdot k]$
(7) Multiplicative identity	$1_\mathbf{F} \cdot v = v$
(8) Left distributivity	$r \cdot [u + v] = [r \cdot u] + [r \cdot v]$
(9) Right distributivity	$[r + s]v = [r \cdot v] + [s \cdot v]$

Definition 4. A **linear subspace** of a linear space V over a field $\mathbf{F}$ consists of a subset $W \subseteq V$ that is also a linear space with the same element 0_V and with the same binary operations $+$ and $\cdot$ already existing for V. □

3.3 Inner Products and Inequalities

Definition 5. An **inner product** (also called a "scalar" or "dot" product) defined on a linear space V over a field $\mathbf{F} \subseteq \mathbf{C}$ is a function

$$\langle\ ,\ \rangle : V \times V \to \mathbf{F}, \quad (v, w) \mapsto \langle v, w \rangle,$$

which satisfies all the properties in **Table 3**, where $\overline{(p, q)} = (p, -q)$ denotes complex conjugation. □

Table 3. Algebraic properties of inner products.

The following properties must hold for all elements $r, s \in \mathbf{F}$ and $u, v, w \in V$.

(1) Nonnegativity of $\langle\ ,\ \rangle$	$\langle v, v \rangle \geq 0_\mathbf{F}$
(2) Positivity of $\langle\ ,\ \rangle$	If $\langle v, v \rangle = 0_\mathbf{F}$ then $v = 0_V$
(3) Linearity of $\langle\ ,\ \rangle$	$\langle (r \cdot u) + (s \cdot v),\ w \rangle = r\langle u, w \rangle + s\langle v, w \rangle$
(4) Anti-symmetry of $\langle\ ,\ \rangle$	$\langle v, w \rangle = \overline{\langle w, v \rangle}$

Theorem 1. (Cauchy-Schwartz Inequality) *For each inner product $\langle\ ,\ \rangle$ on any linear space over any field $\mathbf{F} \subseteq \mathbf{C}$, and for all elements $v, w \in V$,*

$$|\langle v, w \rangle|^2 \leq \langle v, v \rangle \cdot \langle w, w \rangle,$$

with equality if, but only if, v or w equals a multiple of the other: $\mathbf{F}$ contains an element r such that $v = rw$ or $w = rv$.

Definition 6. $\|v\| := \sqrt{\langle v, v\rangle}$

Theorem 2. (Triangle Inequality) *For each inner product* $\langle\,,\,\rangle$ *on any linear space* V *over any field* $\mathbf{F} \subseteq \mathbf{C}$*, and for all elements* $v, w \in V$*,*

$$\|v + w\| \leq \|v\| + \|w\|,$$

with equality if, but only if, v *or* w *equals a nonnegative multiple of the other:* $\mathbf{F}$ *contains an element* $r \geq 0$ *such that* $v = rw$ *or* $w = rv$*.*

Definition 7. A **norm** defined on a linear space V over a field $\mathbf{F} \subseteq \mathbf{C}$ is a function

$$\|\ \| : V \to \mathbf{F}, \quad v \mapsto \|v\|,$$

which satisfies all the properties in **Table 4.**

Table 4. Algebraic properties of norms.

The following properties must hold for all elements $r \in \mathbf{F}$ and $v, w \in V$.

(1) Nonnegativity of $\|\ \|$	$\|v\| \geq 0_{\mathbf{F}}$
(2) Positivity of $\|\ \|$	If $\|v\| = 0_{\mathbf{F}}$ then $v = 0_V$
(3) Homogeneity of $\|\ \|$	$\|rv\| = \|r\| \cdot \|v\|$
(4) Triangle Inequality for $\|\ \|$	$\|v + w\| \leq \|v\| + \|w\|$

Theorem 3. (Reverse Triangle Identity) *For all elements* v *and* w *in a linear space* V *with a norm* $\|\ \|$*,*

$$\|v - w\| \geq \big|\, \|v\| - \|w\| \,\big|.$$

Theorem 4. (Polar Identity) *For all vectors* v *and* w *in a linear space* V *over a field* $\mathbf{F} \subseteq \mathbf{C}$ *with an inner product* $\langle\,,\,\rangle$ *and with the corresponding norm defined by* $\|u\| = \sqrt{\langle u, u\rangle}$*, the following identity holds:*

$$\operatorname{Re}(\langle v, w\rangle) = \frac{1}{4} \cdot \left(\|v + w\|^2 - \|v - w\|^2\right),$$

$$\operatorname{Im}(\langle v, w\rangle) = \frac{i}{4} \cdot \left(\|v + iw\|^2 - \|v - iw\|^2\right).$$

Consequently, if $\mathbf{F} \subseteq \mathbf{R}$*, then*

$$\langle v, w\rangle = \frac{1}{4} \cdot \left(\|v + w\|^2 - \|v - w\|^2\right).$$

Definition 8. In a linear space V with an inner product $\langle\,,\,\rangle$, two vectors v and w are mutually **orthogonal** if, but only if, $\langle v, w\rangle = 0_{\mathbf{F}}$. Also, v and w are mutually **orthonormal** if, but only if, $\langle v, w\rangle = 0_{\mathbf{F}}$ and $\|v\| = 1 = \|w\|$.

Theorem 5. (Pythagorean Theorem) *In a linear space V over a field $\mathbf{F} \subseteq \mathbf{C}$ with an inner product $\langle\ ,\ \rangle$, if two vectors v and w are mutually orthogonal, then $\|v+w\|^2 = \|v\|^2 + \|w\|^2$. Moreover, if $\mathbf{F} \subseteq \mathbf{R}$, then the converse also holds: if $\|v+w\|^2 = \|v\|^2 + \|w\|^2$, then v and w are mutually orthogonal.*

Theorem 6. (Bessel's Inequality) *In a linear space V over a field $\mathbf{F} \subseteq \mathbf{C}$ with an inner product $\langle\ ,\ \rangle$, for each vector $v \in V$ and for each set of nonzero orthogonal vectors $\{w_1, w_2, \dots, w_{m-1}, w_m\} \subseteq V$,*

$$\sum_{k=1}^{m} \frac{|\langle v, w_k\rangle|^2}{\|w_k\|^2} \leq \|v\|^2.$$

3.4 Gram-Schmidt Orthogonalization

The preceding sections have demonstrated the convenience provided by orthogonal or orthonormal bases in the calculation of orthogonal projections. This section describes a method to construct such orthonormal bases.

Theorem 7. (Gram-Schmidt Orthogonalization) *For each linearly independent, finite or infinite, set $S = \{v_1, v_2, \dots, v_k, v_{k+1}, \dots\}$ in a linear space V with an inner product $\langle\ ,\ \rangle$, there exists an orthonormal set $U = \{u_1, u_2, \dots, u_k, u_{k+1}, \dots\}$ such that for each index k,*

$$\langle u_k, v_k\rangle > 0,$$
$$\operatorname{Span}\{u_1, \dots, u_k\} = \operatorname{Span}\{v_1, \dots, v_k\}.$$

Specifically,

$$w_1 := v_1,$$
$$w_{m+1} := v_{m+1} - \sum_{k=1}^{m} \frac{\langle v_{m+1}, w_k\rangle}{\langle w_k, w_k\rangle} \cdot w_k.$$

4. Summary of Orthogonal Projections

In some courses, distributing such a handout as **Exhibit 1** (see next page) may suffice, whereas in other courses longer reviews may prove necessary.

Exhibit 1. Summary of Orthogonal Projections

To state three of the most important results about orthogonal projections, consider a linear subspace $W \subseteq V$ of a linear space V with inner product $\langle \, , \rangle$ and induced norm $\|\vec{\mathbf{x}}\| := \sqrt{\langle \vec{\mathbf{x}}, \vec{\mathbf{x}} \rangle}$. Also, let $\vec{\mathbf{p}} \perp W$ mean $\langle \vec{\mathbf{p}}, \vec{\mathbf{q}} \rangle = 0$ for every $\vec{\mathbf{q}} \in W$.

Theorem A. *If* $\vec{\mathbf{w}} \in W$ *, if* $\vec{\mathbf{v}} \in V$ *, and if* $(\vec{\mathbf{v}} - \vec{\mathbf{w}}) \perp W$ *, then* $\vec{\mathbf{w}}$ *is the member of* W *closest to* $\vec{\mathbf{v}}$ *.*

Proof. For every other member $\vec{\mathbf{u}} \in W$, $\vec{\mathbf{v}} - \vec{\mathbf{w}} \perp \vec{\mathbf{w}} - \vec{\mathbf{u}}$, and the Pythagorean theorem gives

$$\|\vec{\mathbf{v}} - \vec{\mathbf{u}}\|^2 = \|\vec{\mathbf{v}} - \vec{\mathbf{w}}\|^2 + \|\vec{\mathbf{w}} - \vec{\mathbf{u}}\|^2 \geq \|\vec{\mathbf{v}} - \vec{\mathbf{w}}\|^2. \qquad \square$$

Theorem B. *If* $(\vec{\mathbf{w}}_1, \ldots, \vec{\mathbf{w}}_n)$ *is an orthogonal basis for* W *, then for each* $\vec{\mathbf{v}} \in V$ *the element* $\vec{\mathbf{w}} \in W$ *closest to* $\vec{\mathbf{v}}$ *is*

$$\vec{\mathbf{w}} = \sum_{i=1}^{n} \frac{\langle \vec{\mathbf{v}}, \vec{\mathbf{w}}_i \rangle}{\langle \vec{\mathbf{w}}_i, \vec{\mathbf{w}}_i \rangle} \vec{\mathbf{w}}_i.$$

Proof. Verify that $\langle \vec{\mathbf{v}} - \vec{\mathbf{w}}, \vec{\mathbf{w}}_i \rangle = 0$ for each $i \in \{1, \ldots, n\}$, and invoke **Theorem A.** □

Theorem C. (Normal Equations) *If* W *has any basis* $\{\vec{\mathbf{x}}_1, \ldots, \vec{\mathbf{x}}_n\}$ *, then* $\vec{\mathbf{w}} \in W$ *is the element of* W *closest to* $\vec{\mathbf{v}} \in V$ *if, but only if, for each* $i \in \{1, \ldots, n\}$ *,*

$$\langle \vec{\mathbf{w}}, \vec{\mathbf{x}}_i \rangle = \langle \vec{\mathbf{v}}, \vec{\mathbf{x}}_i \rangle.$$

Proof. Verify that $\langle \vec{\mathbf{v}} - \vec{\mathbf{w}}, \vec{\mathbf{x}}_i \rangle = 0$ for each $i \in \{1, \ldots, n\}$, from which we get $\langle \vec{\mathbf{v}} - \vec{\mathbf{w}}, \sum_{i=1}^{n} c_i \vec{\mathbf{x}}_i \rangle = 0$ and $\vec{\mathbf{v}} - \vec{\mathbf{w}} \perp W$, and use **Theorem B.** □

Theorem D. *Let* $V = \mathbf{R}^m$ *and let* $\langle \, , \rangle$ *denote the usual dot product. If* $Q = (\vec{\mathbf{q}}_1, \ldots, \vec{\mathbf{q}}_n) \in \mathbf{M}_{m \times n}(\mathbf{R})$ *is a rectangular matrix with orthonormal columns, and if* $\vec{\mathbf{v}} \in \mathbf{R}^m$ *, then* $Q^T \vec{\mathbf{v}}$ *is the vector of coordinates, with respect to the basis* $(\vec{\mathbf{q}}_1, \ldots, \vec{\mathbf{q}}_n)$ *, of the orthogonal projection of* $\vec{\mathbf{v}}$ *onto the subspace* $W =$ Span $\{\vec{\mathbf{q}}_1, \ldots, \vec{\mathbf{q}}_n\}$.

Proof. Notice that $\left(Q^T \vec{\mathbf{v}}\right)_i = \vec{\mathbf{q}}_i^T \cdot \vec{\mathbf{v}} = \langle \vec{\mathbf{q}}_i, \vec{\mathbf{v}} \rangle$ and apply **Theorem B.** □

In three-dimensional graphics, V models the three-dimensional ambient space, W represents the computer screen, and **Theorem B** describes the image $\vec{w}$ on the screen of a point $\vec{v}$ in space. In statistical data analysis, V is a space of arrays of data, W consists of linear combinations of arrays of design variables, and **Theorem B** provides the linear modeling function closest to the data. In engineering, **Theorem B** shows how to approximate a signal or an image in a space V by a Fourier series, or by wavelets, in W.

In many applications, calculating an orthogonal projection amounts to forming and solving the normal equations described in **Theorem C**, though for some applications such normal equations may exhibit a large sensitivity to rounding errors.

5. Conclusion

Orthogonal projections in abstract linear spaces have applications!

6. Solutions to the Exercises

1. Straightforward calculations of inner products give $p := \langle \vec{x}, \vec{u} \rangle = {}^{2}\!/_{13}$ and $q := \langle \vec{x}, \vec{v} \rangle = -{}^{2663}\!/_{1469}$.

2. $p = {}^{2}\!/_{5} = 0.40$ and $q = {}^{1}\!/_{25} = 0.04$.

3. The OLS line has slope $-{}^{1}\!/_{2}$ and vertical intercept ${}^{9}\!/_{2}$. See Nievergelt [1994].

4. The OLS line has slope ${}^{1}\!/_{2}$ and vertical intercept 5.

5. The OLS line has slope ${}^{9}\!/_{20}$ and vertical intercept ${}^{11}\!/_{5}$. See Nievergelt [1994].

6. a) The normal equations or Gram-Schmidt orthonormalization uses the following inner products:

$$\langle 1, 1 \rangle = \int_{1/8}^{1} 1 \cdot 1 \, dx = {}^{7}\!/_{8},$$

$$\langle x, 1 \rangle = \int_{1/8}^{1} x \cdot 1 \, dx = \left. \frac{x^2}{2} \right|_{1/8}^{1} = \frac{1^2 - (1/8)^2}{2} = \frac{64/64 - 1/64}{2} = \frac{63}{128},$$

$$\langle x, x \rangle = \int_{1/8}^{1} x \cdot x \, dx = \left. \frac{x^3}{3} \right|_{1/8}^{1} = \frac{1^3 - (1/8)^3}{3} = \frac{511}{1536},$$

$$w_1 = v_1 = 1,$$

$$w_2 = v_2 - \frac{\langle v_2, w_1\rangle}{\langle w_1, w_1\rangle} w_1 = x - \frac{\langle x, 1\rangle}{\langle 1, 1\rangle} 1 = x - \frac{63/128}{7/8} 1 = x - \frac{9}{16},$$

$$\langle w_2, w_2\rangle = \int_{1/8}^{1} \left(x - 9/16\right)^2 \, dx = \left. \frac{\left(x - 9/16\right)^3}{3} \right|_{1/8}^{1} = \frac{343}{6144} = \frac{7^3}{2^{11} \times 3},$$

$$u_1 = \frac{1}{\|w_1\|} w_1 = \frac{1}{\sqrt{7/8}} = \frac{2\sqrt{2}}{\sqrt{7}},$$

$$u_2 = \frac{1}{\|w_2\|} w_2 = \frac{\sqrt{6144}}{\sqrt{343}} \left(x - 9/16\right) = \frac{32\sqrt{6}}{7\sqrt{7}} \left(x - 9/16\right).$$

b) Let $h : [1/8, 1] \to \mathbf{R}$ denote the restriction of the cube root to $[1/8, 1]$.

$$\langle h, w_1\rangle = \int_{1/8}^{1} \sqrt[3]{x} \cdot 1 \, dx = \left. \frac{x^{4/3}}{4/3} \right|_{1/8}^{1} = \frac{1 - 1/16}{4/3} = \frac{45}{64} = \frac{3^2 \times 5}{2^6},$$

$$\langle h, w_2\rangle = \int_{1/8}^{1} \sqrt[3]{x} \cdot \left(x - 9/16\right) \, dx = \left. \frac{x^{7/3}}{7/3} \right|_{1/8}^{1} - (9/16)\langle h, w_1\rangle$$

$$= \frac{1 - 1/128}{7/3} - \frac{9 \times 45}{16 \times 64} = \frac{213}{7168} = \frac{3 \times 71}{2^{10} \times 7},$$

$$
\begin{aligned}
q &= \frac{\langle h, w_1\rangle}{\langle w_1, w_1\rangle} w_1 + \frac{\langle h, w_2\rangle}{\langle w_2, w_2\rangle} w_2 \\
&= \frac{45/64}{7/8} 1 + \frac{213/7168}{343/6144} \left(x - 9/16\right) \\
&= \frac{2 \times 3^2 \times 71}{7^4} x + \frac{45}{56} - \frac{213/7168}{343/6144} \frac{9}{16} \\
&= \frac{2 \times 3^2 \times 71}{7^4} x + \frac{3^2 \times 269}{2 \times 7^4} = \frac{1278}{2401} x + \frac{2421}{4802} \\
&\approx 0.532\,278\,2174\, x + 0.504\,164\,9313.
\end{aligned}
$$

Thus,

$$
\begin{aligned}
c_0 &= 2421/4802 \approx 0.504\,164\,9313, \\
c_1 &= 1278/2401 \approx 0.532\,278\,2174.
\end{aligned}
$$

The result appears in **Figure 5**.

c) The affine polynomial $q(x) = c_0 + c_1 x$ just obtained may be used to approximate the cube root $\sqrt[3]{\ }$ on the interval $[1/8, 1]$. For instance, if

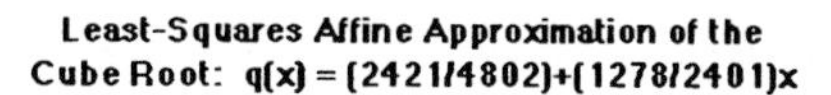

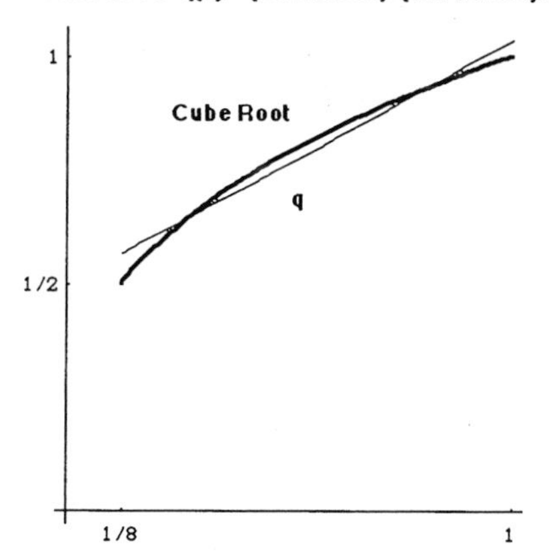

Figure 5. A least-squares affine approximation of the cube root.

$x := {}^{64}\!/_{125}$ then $\sqrt[3]{{}^{64}\!/_{125}} = {}^{4}\!/_{5} = 0.8$ and $q({}^{64}\!/_{125}) = c_0 + c_1\,{}^{64}\!/_{125} \approx 0.776\,691\,3786 \approx 0.8$.

d) The affine polynomial $q(x) = c_0 + c_1 x$ may be used to approximate the cube root $\sqrt[3]{\ }$ on all of $\mathbf{R}$. For instance, if $z := 27$ then $\sqrt[3]{27} = 3$. Moreover, $z = 27 = 64 \times {}^{27}\!/_{64}$ with ${}^{27}\!/_{64} \in [{}^{1}\!/_{8}, 1]$. Hence $q({}^{27}\!/_{64}) = c_0 + c_1\,{}^{27}\!/_{64} \approx 0.728\,719\,8043$ and

$$\begin{aligned}\sqrt[3]{27} &= \sqrt[3]{64 \times {}^{27}\!/_{64}} = \sqrt[3]{64} \times \sqrt[3]{{}^{27}\!/_{64}} \approx \sqrt[3]{64} \cdot q({}^{27}\!/_{64})\\ &\approx 4 \times 0.728\,719\,8043\\ &\approx 2.914\,879\,217.\end{aligned}$$

Thus, a one-digit mini-calculator could evaluate the polynomial q, which requires only one multiplication and one addition, to compute $\sqrt[3]{27}$ to all displayed digits.

e) Let $D(x) := q(x) - \sqrt[3]{x} = c_0 + c_1 x - \sqrt[3]{x}$ denote the discrepancy. For the local extremum of D on $]{}^{1}\!/_{8}, 1[$, apply calculus:

$$\begin{aligned}D(x) &= c_0 + c_1 x - \sqrt[3]{x},\\ D'(x) &= c_1 - ({}^{1}\!/_{3}) \cdot x^{-{}^{2}\!/_{3}},\\ 0 &= c_1 - ({}^{1}\!/_{3}) \cdot x^{-{}^{2}\!/_{3}},\\ x_* &= (3c_1)^{-{}^{3}\!/_{2}} \approx 0.495\,575\,784\,658\ldots,\\ D(x_*) &= c_0 + c_1 x_* - \sqrt[3]{x_*} = c_0 - ({}^{2}\!/_{3})/\sqrt{3c_1} \approx -0.023\,403\ldots,\end{aligned}$$

$$\begin{aligned}D''(x) &= ({}^{2}\!/_{9})x^{-{}^{5}\!/_{3}} > 0,\\ D(1) &= c_0 + c_1 - \sqrt[3]{1} \approx 0.036\,443\ldots,\\ D({}^{1}\!/_{8}) &= c_0 + c_1({}^{1}\!/_{8}) - \sqrt[3]{{}^{1}\!/_{8}} = c_0 + c_1/8 - {}^{1}\!/_{2} \approx 0.070\,699\,708\ldots.\end{aligned}$$

Thus, the maximum absolute discrepancy occurs at $1/8$, where $|D(1/8)| = 0.070\,699\,708\ldots \approx 0.071$. Because the cube root also has its minimum absolute value there, the maximum relative discrepancy also occurs at $1/8$, where

$$\frac{|D(1/8)|}{|\sqrt[3]{1/8}|} = \frac{0.070\,699\,708\ldots}{1/2} = 2 \times 0.070\,699\,708\ldots \approx 0.142.$$

7. a) Use the normal equations or Gram-Schmidt orthonormalization:

$$\begin{aligned}
\langle 1, 1\rangle &= \int_{-1}^{0} 1 \cdot 1\, dx = 1, \\
\langle x, 1\rangle &= \int_{-1}^{0} x \cdot 1\, dx = \left.\frac{x^2}{2}\right|_{-1}^{0} = -\frac{1}{2}, \\
w_2 &= x - \frac{\langle x, 1\rangle}{\langle 1, 1\rangle} \cdot 1 = x - \frac{-1/2}{1} \cdot 1 = x + 1/2, \\
\langle w_2, w_2\rangle &= \int_{-1}^{0} (x + 1/2)^2\, dx = \left.\frac{(x + 1/2)^3}{3}\right|_{-1}^{0} = \frac{1}{12}, \\
\langle \exp_2, 1\rangle &= \int_{-1}^{0} 2^x \cdot 1\, dx = \int_{-1}^{0} e^{x \cdot \ln(2)} \cdot 1\, dx = \left.\frac{1}{\ln(2)} \cdot e^{x \cdot \ln(2)}\right|_{-1}^{0} \\
&= \frac{1}{2\ln(2)}, \\
\langle \exp_2, x\rangle &= \int_{-1}^{0} e^{x \cdot \ln(2)} \cdot x\, dx \\
&= \left. x \cdot \frac{1}{\ln(2)} \cdot e^{x \cdot \ln(2)}\right|_{-1}^{0} - \int_{-1}^{0} \frac{1}{\ln(2)} \cdot e^{x \cdot \ln(2)}\, dx \\
&= \frac{1}{\ln(2)} \cdot e^{-\ln(2)} - \frac{1}{\ln(2)} \cdot \frac{1}{\ln(2)} = \frac{1}{2\ln(2)} \cdot \left(1 - \frac{1}{\ln(2)}\right), \\
\langle \exp_2, w_2\rangle &= \langle \exp_2, x + 1/2\rangle = \langle \exp_2, x\rangle + (1/2)\langle \exp_2, 1\rangle \\
&= \frac{1}{2\ln(2)} \cdot \left(\frac{3}{2} - \frac{1}{\ln(2)}\right), \\
g(x) &= \langle \exp_2, 1\rangle \cdot 1 + \langle \exp_2, w_2\rangle \cdot w_2 \\
&= \frac{1}{2\ln(2)} + \frac{1}{2\ln(2)} \cdot \left(\frac{3}{2} - \frac{1}{\ln(2)}\right) \cdot (x + 1/2) \\
&= \frac{1}{2\ln(2)} \cdot \left(\left[5 - \frac{3}{\ln(2)}\right] + \left[9 - \frac{6}{\ln(2)}\right] \cdot x\right) \\
&\approx 0.969\,368 + 0.496041x.
\end{aligned}$$

The result appears in **Figure 6**.

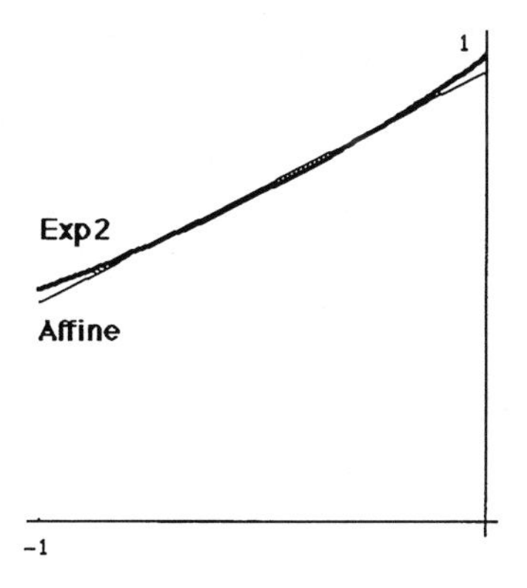

Figure 6. A least-squares approximation of $x \mapsto 2^x$ by an affine polynomial.

b) $$\begin{aligned} \exp_2(-1/2) &= 2^{-1/2} = 1/\sqrt{2} &&= 0.707\,106\,781\,188\ldots \\ g(-1/2) &\approx 0.969\,368 + 0.496041 \times (-1/2) &&\approx 0.721\,348\ldots \end{aligned}$$

c) For each $z \in \mathbf{R}$,

$$e^z = e^{[\ln(2) z/\ln(2)]} = 2^{[z/\ln(2)]} = \exp_2(z/\ln(2)).$$

To transform $z/\ln(2)$ into a number in $[-1, 0]$, define $N := \lceil z/\ln(2) \rceil$ and $x := \{z/\ln(2)\} - N \in]-1, 0]$. Hence, $z/\ln(2) = N + x$, whence

$$e^z = \exp_2(z/\ln(2)) = 2^{[z/\ln(2)]} = 2^{(N+x)} = 2^N \cdot 2^x \approx 2^N \cdot g(x).$$

For example, with $z := 1$,

$$\begin{aligned} z &= 1, \\ z/\ln(2) &= 1/\ln(2) = 1.442\,695\ldots, \\ N &= \lceil z/\ln(2) \rceil = 2, \\ x &= \{z/\ln(2)\} - N = 1.442\,695\ldots - 2 = -0.557\,305\ldots, \\ g(x) &= g(-0.557\,305\ldots) \approx 0.969\,368 + 0.496041 \times (-0.557\,305) \\ &\approx 0.692\,922\ldots, \\ 2^N g(x) &= 2^2 \times 0.692\,922\ldots \\ &\approx 2.771\,688\ldots, \\ e^1 &= 2.718\,281\,828\,46\ldots. \end{aligned}$$

d) Let $D(x) := g(x) - \exp_2(x) = c_0 + c_1 x - 2^x$ denote the discrepancy. For the local extremum of D on $]-1, 0[$, apply calculus:

$$
\begin{aligned}
D(x) &= c_0 + c_1 x - 2^x = c_0 + c_1 x - e^{\ln(2)\cdot x},\\
D'(x) &= c_1 - \ln(2)\cdot e^{\ln(2)\cdot x},\\
0 &= c_1 - \ln(2)\cdot e^{\ln(2)\cdot x},\\
e^{\ln(2)\cdot x} &= c_1/\ln(2),\\
\ln(2)\cdot x &= \ln[c_1/\ln(2)],\\
x_* &= \ln[c_1/\ln(2)]/\ln(2) \approx -0.482\,701\ldots,\\
D(x_*) &= c_0 + c_1 x_* - 2^{x_*} = c_0 + c_1 x_* - e^{\ln(2)\cdot x_*} \approx 0.014\,292\ldots,\\
D''(x) &= -[\ln(2)]^2\cdot e^{\ln(2)\cdot x} < 0,\\
D(-1) &\approx 0.969\,368 + 0.496041\times(-1) - 2^{-1} \approx -0.030\,632\ldots,\\
D(0) &\approx 0.969\,368 + 0.496041\times(0) - 2^{-0} \approx -0.026\,673\ldots.
\end{aligned}
$$

Thus, the maximum absolute discrepancy occurs at -1, where $|D(-1)| = 0.030\,632\ldots \approx 0.031$. Because $\exp_2$ also has its minimum absolute value there, the maximum relative discrepancy also occurs at -1, where

$$\frac{|D(-1)|}{|\exp_2(-1)|} = \frac{0.030\,632\ldots}{1/2} = 2\times 0.030\,632\ldots \approx 0.062.$$

8. See **Figure 7**, top: $b_{f,k} = 0$, $a_{f,0} = 0$, $a_{f,k} = -4[\sin(k\pi/2)]/(k\pi)$;

$$f(x) \approx \frac{4}{\pi}\left[-\cos(x) + \frac{1}{3}\cos(3x) - \frac{1}{5}\cos(5x) + \frac{1}{7}\cos(7x) - \frac{1}{9}\cos(9x) + \cdots\right].$$

9. See **Figure 7**, middle: $a_{g,k} = 0$, $b_{g,k} = -4[\sin(k\pi/2)]/(k^2\pi)$;

$$g(x) \approx \frac{4}{\pi}\left[-\sin(x) + \frac{1}{9}\sin(3x) - \frac{1}{25}\sin(5x) + \frac{1}{49}\sin(7x) - \frac{1}{81}\sin(9x) + \cdots\right].$$

10. See **Figure 7**, bottom: $b_{h,k} = 0$, $a_{h,0} = \pi^2/8$, $a_{h,k} = 4[\sin(k\pi/2)]/(k^3\pi)$;

$$h(x) \approx \frac{\pi^2}{8} + \frac{4}{\pi}\left[\cos(x) - \frac{1}{27}\cos(3x) + \frac{1}{125}\cos(5x) - \frac{1}{343}\cos(7x) + \frac{1}{729}\cos(9x) + \cdots\right].$$

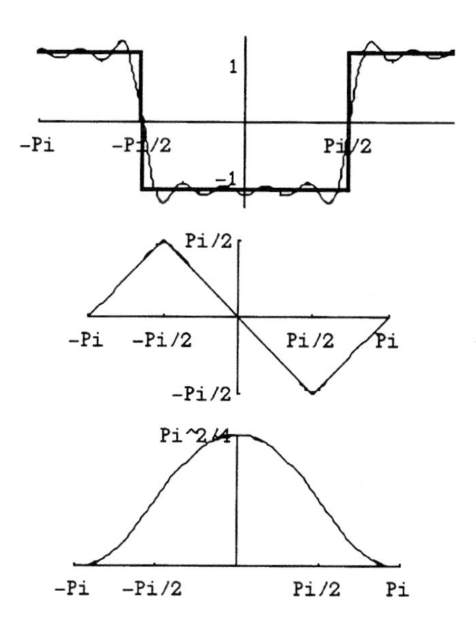

Figure 7. Superpositions of a function and a partial sum of its Fourier series, using 10 terms, demonstrate a faster convergence for smoother functions.

Top:	Piecewise constant.
Middle:	Piecewise affine.
Bottom:	Piecewise quadratic.

11.

$$-1 = f(0) = \frac{4}{\pi}\left[-\cos(0) + \frac{1}{3}\cos(0) - \frac{1}{5}\cos(0) + \frac{1}{7}\cos(0) - \frac{1}{9}\cos(0) + \cdots\right].$$

12.

$$\frac{\pi^2}{4} = h(0) = \frac{\pi^2}{8} + \frac{4}{\pi}\left[\cos(0) - \frac{1}{27}\cos(0) + \frac{1}{125}\cos(0) - \frac{1}{343}\cos(0) + \frac{1}{729}\cos(0) + \cdots\right].$$

References

Hanes, Kit. 1990. *An Introduction to Analytic Projective Geometry and Its Applications.* UMAP Modules in Undergraduate Mathematics and Its Applications: Module 710, Lexington, MA: COMAP, 1990.

Kincaid, David R., and E. Ward Cheney. 1991. *Numerical Analysis: The Mathematics of Scientific Computing,* 1996. 2nd ed. Pacific Grove, CA: Brooks/Cole.

Körner, Thomas William. 1988. *Fourier Analysis.* Cambridge, UK: Cambridge University Press.

________ . 1993. *Exercises for Fourier Analysis.* Cambridge, UK: Cambridge University Press.

Marti, J.T., translated by J.R. Whiteman. 1986. *Introduction to Sobolev Spaces and Finite Element Solution of Elliptic Boundary Value Problems.* Orlando, FL: Academic Press.

Nievergelt, Yves. 1989. *Mathematics in Business Administration.* Homewood, IL: Irwin.

________ . 1992a. *3-D Graphics in Calculus and Linear Algebra.* UMAP Modules in Undergraduate Mathematics and Its Applications: Module 717. Lexington, MA: COMAP, 1992. Reprinted in *UMAP Modules: Tools for Teaching 1991*, edited by Paul J. Campbell, 125–169. Lexington, MA: COMAP, 1992.

________ . 1992b. *Splines in Single and Multivariable Calculus.* UMAP Modules in Undergraduate Mathematics and Its Applications: Module 718. Lexington, MA: COMAP, 1992. Reprinted in *UMAP Modules: Tools for Teaching 1992*, edited by Paul J. Campbell, 39–101. Lexington, MA: COMAP, 1993.

________ . 1994. Total least squares: State-of-the-art regression in numerical analysis. *SIAM Review* 36 (2): 258–264.

Pulskamp, Richard J., and James A. Delaney. 1992. *Computer and Calculator Computation of Elementary Functions.* UMAP Modules in Undergraduate Mathematics and Its Applications: Module 708. Lexington, MA: COMAP, 1991. Reprinted in *The UMAP Journal* 12 (1991): 315–348. Reprinted in *UMAP Modules: Tools for Teaching 1991*, edited by Paul J. Campbell, 1–34. Lexington, MA: COMAP, 1992.

Taylor, Angus E. 1958. *Introduction to Functional Analysis.* New York, NY: Wiley.

About the Author

Yves Nievergelt graduated in mathematics from the École Polytechnique Fédérale de Lausanne (Switzerland) in 1976, with concentrations in functional and numerical analysis of PDEs. He obtained a Ph.D. from the University of Washington in 1984, with a dissertation in several complex variables under the guidance of James R. King. He now teaches complex and numerical analysis at Eastern Washington University.

Prof. Nievergelt is an associate editor of *The UMAP Journal*. He is the author of several UMAP Modules, a bibliography of case studies of applications of lower-division mathematics (*The UMAP Journal* 6 (2) (1985): 37–56), and *Mathematics in Business Administration* (Irwin, 1989).

UMAP

Modules in Undergraduate Mathematics and Its Applications

Published in cooperation with

The Society for Industrial and Applied Mathematics,

The Mathematical Association of America,

The National Council of Teachers of Mathematics,

The American Mathematical Association of Two-Year Colleges,

The Institute for Operations Research and the Management Sciences, and

The American Statistical Association.

Module 762

Optimal Foraging Theory

Steven Kolmes
Kevin Mitchell
James Ryan

Applications of Calculus to Ecology and Biology

COMAP, Inc., Suite 210, 57 Bedford Street, Lexington, MA 02173 (617) 862–7878

INTERMODULAR DESCRIPTION SHEET: UMAP Unit 762

TITLE: Optimal Foraging Theory

AUTHOR:

Steven Kolmes
Dept. of Biology
University of Portland
Portland, OR 97203
kolmes@uofport.edu

Kevin Mitchell
Dept. of Mathematics and Computer Science
Hobart and William Smith Colleges
Geneva, NY 14456
mitchell@hws.edu

James Ryan
Dept. of Biology
Hobart and William Smith Colleges
Geneva, NY 14456
ryan@hws.edu

MATHEMATICAL FIELD: Calculus

APPLICATION FIELD: Ecology, biology

TARGET AUDIENCE: Students in an elementary mathematical modeling course or in a first-semester calculus course

ABSTRACT: This Unit is an introduction to the modeling of animal foraging behavior. We look first at how such basic factors as search and handling times and energy content may affect prey choice by foraging animals. We construct an elementary model that maximizes the net energy intake rate of the forager and that can predict "optimal" diets of animals. We then modify and refine this model to take into account factors such as prey recognition time, food patches, and central place foraging. Data from several experiments illustrate all the models.

PREREQUISITES: Differential calculus

RELATED UNITS: Unit 688: *Time Resources in Animals*, by Kevin Mitchell and Steven Kolmes. Reprinted in *UMAP Modules: Tools for Teaching 1988*, 83–143. Arlington, MA: COMAP, 1989.

COMAP, Inc., Suite 210, 57 Bedford Street, Lexington, MA 02173
(800) 77–COMAP = (800) 772–6627 (781) 862–7878

Optimal Foraging Theory

Steven Kolmes
Dept. of Biology
University of Portland
Portland, OR 97203
kolmes@uofport.edu

Kevin Mitchell
Dept. of Mathematics and Computer Science
Hobart and William Smith Colleges
Geneva, NY 14456
mitchell@hws.edu

James Ryan
Dept. of Biology
Hobart and William Smith Colleges
Geneva, NY 14456
ryan@hws.edu

Table of Contents

Modules and Monographs in Undergraduate Mathematics and its Applications (UMAP) Project

The goal of UMAP is to develop, through a community of users and developers, a system of instructional modules in undergraduate mathematics and its applications, to be used to supplement existing courses and from which complete courses may eventually be built.

The Project was guided by a National Advisory Board of mathematicians, scientists, and educators. UMAP was funded by a grant from the National Science Foundation and now is supported by the Consortium for Mathematics and Its Applications (COMAP), Inc., a nonprofit corporation engaged in research and development in mathematics education.

1. Introduction

The basic premise of this Module is that all animals can be considered to be predators. Lions eat zebra and wildebeest; sparrows capture seeds. The distinction between carnivores (meat eaters) and herbivores (vegetation eaters) is ignored here because there are fundamental attributes shared by all organisms that search for food. An animal searching for prey must make certain key choices, including:

- what type or types of prey to search for in the first place;
- when to leave an area in which it has already been searching for prey and move on to a new area.

We will consider such choices and develop "rules" that an animal should follow when foraging to maximize energy gain.

Of course, maximum benefit means maximum fitness in terms of survival and overall reproductive success. Behavioral ecologists began using optimization models to address feeding strategies in the early 1970s. This approach, now known as *optimal foraging theory*, tries to understand why particular foraging behaviors are adaptive. Foraging is viewed in terms of the trade-off between foraging costs and benefits, and researchers seek to determine if organisms are maximizing their net benefits.

The general assumption underlying all of the work in this Module is that an animal should behave so as to maximize the *net* energy intake per unit time. Several factors help determine this intake rate. Every type of searching behavior expends a certain amount of energy (e.g., standing in line at a cafeteria or chasing an ice cream truck). Every type of prey has a certain abundance in the environment, and it will generally take less time to encounter a common prey item than a rare one. Every type of prey has its own nutritional value (carrot sticks versus a bucket of chicken wings) and its own handling time and effort (it is hard to open live oysters, easy to open a cardboard hamburger container). The entire constellation of these choices should act together to shape an overall feeding strategy that an animal would find advantageous in terms of net energy acquisition.

Another basic assumption of this Module is that natural selection acts over time to maximize the fitness of organisms, and that an organism's behavioral repertoire in terms of food selection is an important component contributing to this fitness. Foraging models are attractive vehicles to test evolutionary theory because the components of benefits and costs are often simple to articulate and measure.

We begin this Module by defining basic terms, such as handling time, encounter rate, search time, and profitability, in order to set the stage for the models that follow. The first model is a simple prey-choice model that maximizes the net energy intake rate given prey with different profitabilities. A series of examples and problems follows and expands this simple model.

Next, we deal with the problem of prey recognition. Recognizing prey as edible and assessing their "profitability" is an important component of foraging for many predators.

Once we are comfortable with how search time, handling time, and recognition time affect the model, we consider variation in patch quality. Prey are not distributed uniformly in the environment and as a result some patches are better than others. Here we take up the marginal value theorem, which answers the question of how long a predator should stay in a particular patch. We then consider an extension of these ideas known as *central place foraging*, where a round trip cost is applied to parents provisioning young at the nest.

We conclude the Module with some of the complications and criticisms of these optimization models, including the problem of acquiring vital nutrients and hunting in social groups.

2. Building Intuition

2.1 Basic Definitions

For any predator, each prey type[1] will have its own *handling time*, denoted by h. This is the amount of time that it should take a predator, once it has located a prey item, to subdue it, perhaps to break open its shell, etc., and to consume it. Search time, the time spent locating the item, is not included.

Each prey type has its own *net caloric value*, denoted by E. This is the amount of energy that can be extracted from the prey when the prey is digested minus the amount of energy expended in breaking the prey open, chewing it up, and digesting it.

The *profitability* of a prey type, denoted by P, is the net caloric value of the prey divided by the handling time. That is, $P = E/h$, with typical units being calories per minute or joules per second.

Most predators live in environments containing a mixture of potential prey items. A predator should select a group of prey items so that their individual profitabilities and their population densities combine to yield the greatest overall energetic gain. *Prey density*, the number of individuals of each prey type per unit of area in the habitat, matters because a predator that eats only the most individually profitable prey items will spend more time searching for something acceptable to eat than would a predator that ate every type of prey it encountered. This Module is concerned with understanding the trade-off between individual profitability and individual prey density that should produce an overall advantageous diet breadth in a given situation.

It is convenient to translate the notion of prey density to a prey *encounter rate*, λ. The more dense a particular prey type is, the higher the encounter rate

[1]The word "type" is meant to be quite general. For one predator, "type" may mean entirely different prey (a robin feeding on different insects). For another predator, "type" might mean different sizes of the same prey (a crab feeding on different size mussels).

will be. In particular, the encounter rate for prey type i is

$$\lambda_i = \frac{\text{total encounters with type } i}{\text{total search time}}.$$

For example, a squirrel in an oak tree might encounter 25 acorns in a 10 minute interval, yielding an encounter rate for acorns of $\lambda = 25/10 = 2.5$ acorns per minute.

An additional complication is that prey items can be hard to identify. Later we will make the concept of handling time more realistic by subdividing it into a handling time in the strict sense and a recognition time.

Exercises

1. Suppose that a forager encounters two prey types: A and B. On average, type A is encountered once every k seconds.

- **a)** What is the encounter rate λ_A for items of type A?
- **b)** Suppose that there are *twice as many* prey of type B. What is the encounter rate λ_B for items of type B?
- **c)** How often will the combined prey types be encountered?

2. Imagine that you are walking down an avenue in New York City. Every six blocks you might expect to pass a kosher delicatessen serving great pastrami sandwiches. Every three blocks you might pass a knish cart serving decent potato knishes. Every block there will be a hotdog cart serving pretty mediocre hotdogs. Assume a walking speed of 1 block per 2 minutes.

- **a)** What would the individual encounter rates be for delicatessens, for knishes, and for hotdogs? (Answers should have units of encounters per minute.)
- **b)** Under the same circumstances, what would the encounter rate be for all items combined?
- **c)** If you can't afford a deli, what would your encounter rate be for both types of food carts combined?
- **d)** How could you change your behavior to raise your encounter rate for delicatessens to the same level that you calculated for hotdog carts in the first part? Would there be any additional costs associated with this new behavior? What would this do to your knish and hotdog encounter rates?

2.2 Search Time

As we have already noted, for optimal foraging a predator should select a group of prey types in such a way that their profitability and encounter rates combine to maximize the energetic gain from foraging. The encounter rates of the various prey types are crucial because they influence the search time involved in foraging.

It is easy to see how search time changes as a function of the number of types of prey included in the diet. Assume that it takes a predator k seconds to locate or encounter a particular prey type. If we now double the number of prey types searched for and we assume equal densities of both prey types, we expect roughly twice as many encounters in the same amount of time. That is, the search time should be roughly halved to $k/2$ seconds. In the same way, if a predator is searching for any one of three prey types, all other things being equal, we expect the search time to drop to one-third of the time required to locate a single prey type. In other words, the search time is inversely proportional to the number of prey types in the diet (**Figure 1a**).

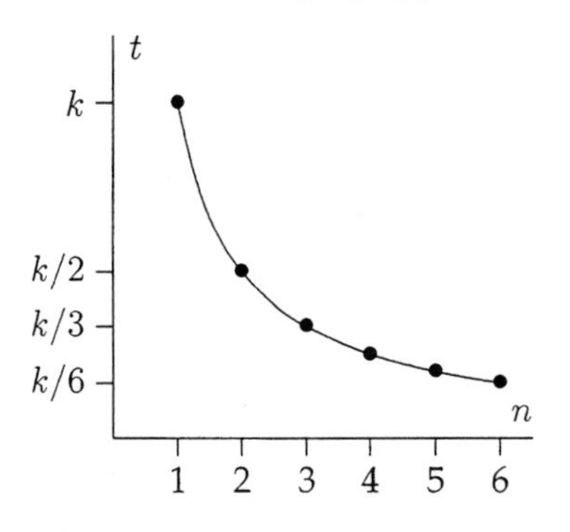

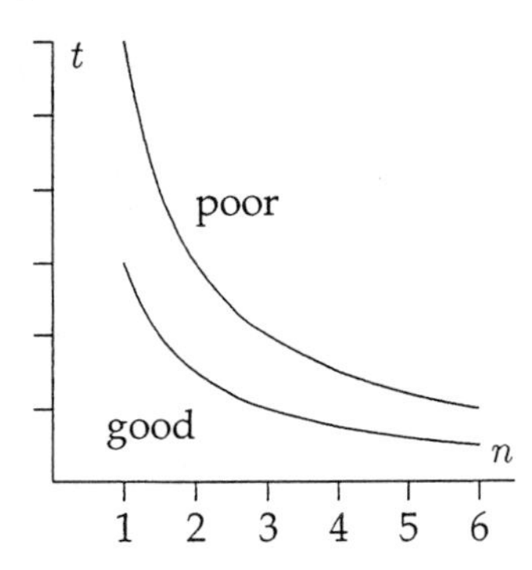

Figure 1. a. (left) The inverse relationship between search time t and the number of prey types n, in the diet of a predator. **b.** (right) Search time in good and poor habitats contrasted.

This inverse relationship means that when the number of prey types n is large, then search time t is small, and vice versa. Further, the graph of search time is steep when n is small but flattens out quickly as n becomes large. This feature has a critical interpretation for our model. The steepness for small values of n means that search time can be dramatically decreased by a small increase in the number of prey types in the diet. Thus, a predator cannot be overly fussy about prey (i.e., search for only one or two types) without expending enormous amounts of time (and energy) searching. The flatness of the curve for larger n means that search time will not be changed significantly by small increases in the number of prey types in this situation. The difference in search times between hunting for 10 or 11 types of prey is quite small. After a certain point, it barely pays to include additional prey types in the diet.

The analysis so far has been independent of the type of habitat in which the predator is located. Search time decreases as the number of prey types included in the diet increases. Thus, if a predator is located in a good habitat where prey of all types is bountiful, then this increased density of each type of prey lowers the search time proportionally for each prey type. Conversely, if prey is less plentiful in a poor habitat, the search time will increase proportionally. Notice the increased steepness for small values of n in the poor habitat graph (**Figure 1b**). Here, especially, search time can be dramatically altered by increasing the number of prey types hunted. This strategy is less important in terms of the number of seconds saved in better habitats.

2.3 Profitability

Recall that profitability P equals E/h, the net caloric value of the prey divided by the handling time. Predators are interested in highly profitable prey.[2] In general, a strategy of selecting prey with low handling times will not be successful since easily eaten prey may be quite small and of low net caloric value. Nor does a strategy of eating only items of high net caloric value work since they may also have large handling times. It usually requires some sort of balancing to achieve the maximum profitability.

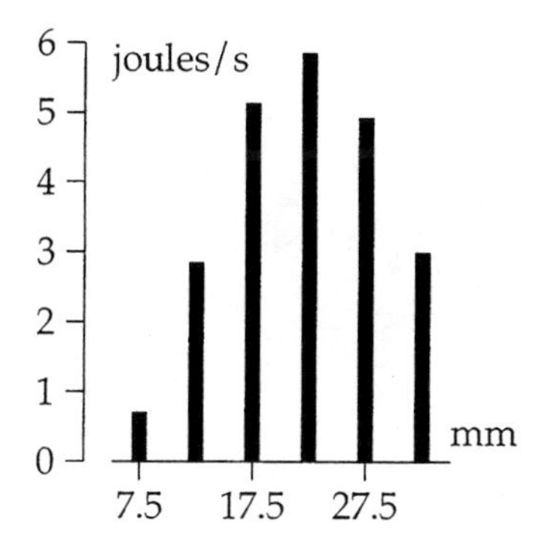

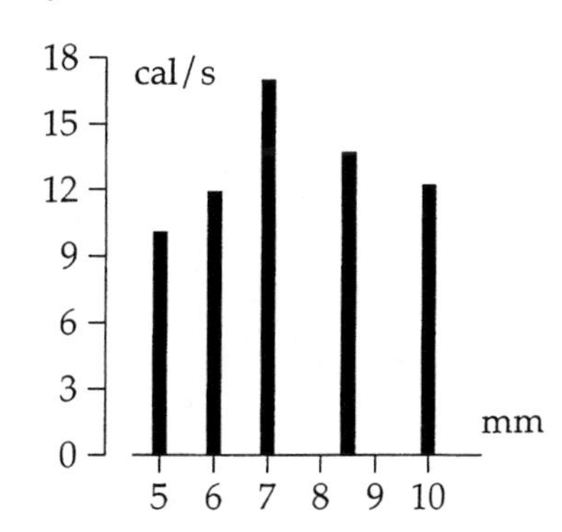

Figure 2. The profitability of various sized mussels (**a**, left) and flies (**b**, right). See Elner and Hughes [1978] and Davies [1977].

As a first example, consider crabs eating mussels of various sizes [Elner and Hughes 1978], illustrated in **Figure 2a**. Smaller mussels are more easily cracked and eaten than large ones but yield little net caloric value and so have low profitability. While larger mussels have a high net caloric value, the substantial handling time involved in eating them makes their profitability low for the crab. The most appropriate strategy for the crab is to choose mussels of a moderate size that yield a reasonable net caloric value without a large investment of time.

A similar example is provided by pied wagtails feeding on flies of various sizes [Davies 1977], illustrated in **Figure 2b**. The most profitable flies are not the largest, because those require greater handling times. Again, the moderate sized prey is the most "economical," providing the highest net caloric intake per unit handling time.

---

[2]Dieters, on the contrary, try to minimize profitability by eating things like celery that have low net caloric value and relatively high handling (chewing) time. A sense of "fullness," in humans at least, is based in part upon the physical stimulus of chewing.

Exercise

3. In a laboratory setting, the size of prey items can be controlled and the other factors that affect handling time can be examined. The white-throated savanna monitor, *Varanus albigularis*, is a large lizard inhabiting the arid savanna of southern Africa. In a study [Kaufman et al. 1996] of their prey choice, 16 lizards were presented with a variety of diets of California land snails, *Helix aspersa* and *H. asperta*.
   a) As an indirect measure of the net energy content of the prey, the mass of the snail was used. What assumptions did researchers make in using this estimate? What are the advantages and disadvantages of such an estimate?
   b) In one experiments, three types of whole large snails (11.1 ± 0.2 g) were used. Lizards were presented with a choice of: (1) snails that were *intact* with undamaged shells, (2) snails that had *crushed* shells, and (3) snails that were completely *shelled*. The handling times for three types were 16.33 s, 6.40 s, and 1.32 s, respectively. Determine the profitabilities of each type.
   c) In another experiment, lizards were presented with large (11.1 ± 0.2 g) and small (2.9 ± 0.1 g) intact snails. Kaufman et al. report that the profitabilities were 0.64 g/s and 0.27 g/s, respectively. What were the handling times for each type?

3. A Simple Model of Foraging

So far we have treated profitability, search times, and caloric intake rates rather intuitively. For example, we treated the density or encounter rates of all prey types as equal. This is seldom the case in a natural environment, though it can be "arranged" in laboratory experiments. Fortunately, even when prey densities vary, the analysis is still fairly simple to carry out.

First, we must specify exactly which "currency" or quantity the foraging model we are building will optimize. Recall that the premise of the model is that predators act to maximize their net caloric intake rate, E/T. Here E represents the net caloric value of the items eaten and T is the total foraging time (both search and handling time). This model ignores certain features, such as a forager's need to avoid its own predators. Other "optimal" premises are possible. For example, we might construct a model whose premise was to maximize the foraging efficiency, that is, maximize the energy gained over the energy expended.

3.1 The Algebra of Prey Choice

To understand how the net caloric intake rate is maximized, consider a predator hunting for only two types of prey. The densities of the prey may be different, so the two prey are encountered at rates λ_1 and λ_2 prey per minute. If the predator is *unselective* in its searching (does not seek out one prey while ignoring the other), then in t minutes of search time it will encounter $\lambda_1 t$ prey of type 1 and $\lambda_2 t$ prey of type 2. Let E_1 and E_2 denote the net caloric values of the two prey types. Then the prey encountered in t minutes of unselective searching have a net caloric value of

$$E = \lambda_1 t E_1 + \lambda_2 t E_2 = t(\lambda_1 E_1 + \lambda_2 E_2).$$

Next, let h_1 and h_2 denote the handling times for each prey type. The *total foraging time, T,* is the time required to search for and eat the aforementioned prey. It will be the sum of the search time plus the individual handling times for the encountered numbers of the two prey:

$$T = t + (\lambda_1 t)h_1 + (\lambda_2 t)h_2 = t(1 + \lambda_1 h_1 + \lambda_2 h_2).$$

Thus, the overall net caloric intake rate of this unselective diet is

$$\frac{E}{T} = \frac{t(\lambda_1 E_1 + \lambda_2 E_2)}{t(1 + \lambda_1 h_1 + \lambda_2 h_2)} = \frac{\lambda_1 E_1 + \lambda_2 E_2}{1 + \lambda_1 h_1 + \lambda_2 h_2}.$$

Notice that search time does not appear as a factor in the final expression for the net caloric intake rate.

An entirely analogous argument shows that if only one prey type is hunted, then the overall net caloric intake rate will be

$$\frac{E}{T} = \frac{\lambda_1 E_1}{1 + \lambda_1 h_1}.$$

More generally, if n different prey types are involved in the diet, each with its own encounter rate λ_i, net caloric value E_i, and handling time h_i, then unselective predation yields a net caloric intake rate of

$$\frac{E}{T} = \frac{\lambda_1 E_1 + \lambda_2 E_2 + \cdots + \lambda_n E_n}{1 + \lambda_1 h_1 + \lambda_2 h_2 + \cdots + \lambda_n h_n}.$$

The optimal breadth of diet for each predator occurs when the intake rate, E/T, is maximized. The expression just derived for E/T can be used to determine exactly which prey types ought to be included in the diet. The answer turns out to depend on the following fact of algebra.

Fact 1. *Let A, B, a, and b be any positive real numbers. Then*

$$\frac{A}{B} < \frac{A+a}{B+b} \iff \frac{A}{B} < \frac{a}{b}.$$

Proof: It is straightforward to check that

$$\frac{A}{B} < \frac{A+a}{B+b} \iff AB + Ab < AB + aB \iff Ab < aB \iff \frac{A}{B} < \frac{a}{b}. \qquad \square$$

We can use this fact to determine the optimal diet for a forager. Continuing with the example of the predator with two potential prey types, assume that the profitability for the first type is greater than the profitability of the second, that is, $E_1/h_1 > E_2/h_2$. Which diet maximizes net caloric intake: a diet of only prey type 1 or a diet of both prey types? To answer the question, compare the rates of intake for each diet and ask under what conditions the second diet yields the greater intake rate. Using the expressions for both rates of intake developed earlier and **Fact 1** (with $A = \lambda_1 E_1$, $B = 1 + \lambda_1 h_1$, $a = \lambda_2 E_2$, and $b = \lambda_2 h_2$), we see that

$$\frac{\lambda_1 E_1}{1+\lambda_1 h_1} < \frac{\lambda_1 E_1 + \lambda_2 E_2}{1+\lambda_1 h_1 + \lambda_2 h_2} \iff \frac{\lambda_1 E_1}{1+\lambda_1 h_1} < \frac{\lambda_2 E_2}{\lambda_2 h_2} = \frac{E_2}{h_2}.$$

That is, prey type 2 should be included in the diet only when its profitability, E_2/h_2, is greater than the net caloric intake rate for prey type 1, which is $\lambda_1 E_1/(1+\lambda_1 h_1)$. Equivalently, type 2 should be excluded from the diet when it is less profitable than the intake rate from a diet of type 1. Notice that the encounter rate for prey type 2 is not important; only its profitability is of concern in this decision.

Suppose that there were three potential prey types ranked by profitability: $E_1/h_1 > E_2/h_2 > E_3/h_3$. Of course, the most profitable prey type is included in the diet. Using the above criterion, we can tell whether the predator should include prey type 2 in its diet. If prey type 2 should be excluded, then so should prey type 3, since it is even less profitable than type 2. But if type 2 is included in the diet, then we must determine if type 3 is to be included or excluded. Exactly the same analysis as above must be carried out. When is the intake rate for the first two items less than for all three items?

$$\frac{\lambda_1 E_1 + \lambda_2 E_2}{1+\lambda_1 h_1 + \lambda_2 h_2} < \frac{\lambda_1 E_1 + \lambda_2 E_2 + \lambda_3 E_3}{1+\lambda_1 h_1 + \lambda_2 h_2 + \lambda_3 h_3} \iff \frac{\lambda_1 E_1 + \lambda_2 E_2}{1+\lambda_1 h_1 + \lambda_2 h_2} < \frac{E_3}{h_3}.$$

A prey item is to be included in the diet if and only if its profitability is greater than the net caloric intake rate of the diet without the item.

Exercise

4. **Fact 1** is used to prove the inequality comparing the net caloric intake rates of the two and three item diets above. Identify A, B, a, and b in this inequality.

5. In a field study of the foraging ecology of the swallow (*Hirundo rustica*), Turner [1982] estimated the net caloric value and handling times of large insect prey to be $E_L = 32.9$ calories and $h_L = 9.1$ seconds, respectively. For small insect prey, the net caloric value and handling times were $E_S = 5.3$ calories and $h_S = 3.2$ seconds, respectively.

   a) Show that large insect prey are more profitable than small.

   b) Calculate the minimal encounter rate λ_L required in order for a swallow's optimal diet to consist of large insect prey only.

3.2 Rules for Optimal Foraging

More generally, the discussion above shows that if the potential diet consists of n prey types ranked by profitability, then the optimal forager should:

1. Always include the most profitable prey type in its diet. This is true even if the item is so rare that it cannot contribute a great deal to the total net caloric intake.

2. Beginning with the individually most profitable type of prey, continue to include prey types in its diet until the profitability of a type falls below the net caloric intake rate of the diet *without* that type. Prey type $k+1$ is to be included in the diet if and only if

$$\frac{\lambda_1 E_1 + \lambda_2 E_2 + \cdots + \lambda_k E_k}{1 + \lambda_1 h_1 + \lambda_2 h_2 + \cdots + \lambda_k h_k} < \frac{E_{k+1}}{h_{k+1}}.$$

3. Include or exclude a prey type in the diet regardless of that prey type's encounter rate (λ_{k+1} does not appear in the equation above), but with consideration of the encounter rates of the types previously included in the diet.

4. Exclude low-ranking prey as an "all or nothing" affair depending on the inequality at step 2.

These rules provide a way to test whether a predator is an optimal forager.

Example: A Diet for Blue Jays. In this example, we consider a hypothetical case of blue jays eating three possible prey: moths, worms, and grubs (**Table 1**). What is the optimal diet using the rules outlined above?

Table 1.

Items in a jay's hypothetical diet.

Item	E kcal	h min	$P = E/h$ kcal/min	λ #/min
worm	162	3.6	45	0.2
moth	24	0.6	40	3.0
grub	40	1.6	25	3.0

Solution. Worms have the highest profitability, so they are included in the diet (Rule 1). The net intake rate for a diet of worms only is

$$\frac{E}{T} = \frac{0.2(162)}{1 + 0.2(3.6)} \approx 18.8 \text{ kcal/min}.$$

The profitability of moths (40 kcal/min) is greater than the net intake rate of the diet of worms. Rule 2 says to include moths in the diet. The intake rate for a diet of worms and moths is

$$\frac{E}{T} = \frac{0.2(162) + 3.0(24)}{1 + 0.2(3.6) + 3.0(0.6)} \approx 29.7 \text{ kcal/min.}$$

The profitability of grubs (25 kcal/min) is lower than the intake rate from the diet of worms and moths, so grubs are not be included in the diet. The worms and moths together provide the optimal diet and the net intake rate is 29.7 kcal/min. If jays, under these conditions, were observed to eat only worms and moths, that would be supporting evidence for the model.

Exercises

6. A forager has prey types A through D available. Each row in **Table 2** shows a diet for the animal. For example, in Diet 1, the animal eats all of the type A and C items it encounters, and only one-fourth of the type B and D items. Which of these diets are *not* possible under the assumption that *the animal is foraging optimally*? Explain.

Table 2.

Data for **Exercise 6**.

	Item A	Item B	Item C	Item D
Diet 1	all	one-fourth	all	one-fourth
Diet 2	all	none	none	all
Diet 3	all	all	none	all
Diet 4	one-fourth	one-fourth	one-fourth	one-fourth
Diet 5	one-half	one-half	none	none

7. Consider the following hypothetical situation. A gull is searching for food at the shore. Mussels are very common ($\lambda_m = 10/\text{min}$), contain a reasonable caloric reward ($E_m = 50$ kcal), but are hard to pry open ($h_m = 2$ min). Shore worms are less common ($\lambda_w = 1/\text{min}$), contain fewer calories ($E_w = 30$ kcal), but are very easy to eat ($h_w = 0.25$ min). Crabs are uncommon ($\lambda_c = 0.5/\text{min}$), full of good food ($E_c = 180$ kcal), but are hard to subdue ($h_c = 3$ min).

a) Put this information in tabular form (as in **Table 1**) and calculate the profitabilities for each item.

b) Determine the optimal diet and the corresponding net caloric intake rate.

c) Suppose, instead, that the mussels were divided into two size categories that were easy to tell apart: small mussels ($E_{sm} = 20$ kcal) and large mussels ($E_{lm} = 200$ kcal), with the same encounter rate ($\lambda_{sm} = \lambda_{lm} = 10/\text{min}$) and the same handling time ($h_{sm} = h_{lm} = 2$ min) for both. Assume that shore worms and crabs remain available, as above. Calculate the new optimal diet.

8. a) Rule 2 for optimal foraging says that prey type $k+1$ is to be included in the diet if its profitability E_{k+1}/h_{k+1} exceeds the net caloric intake rate for the diet *without* that type. Show that once this item is included in the diet that E_{k+1}/h_{k+1} *still exceeds* the new net intake rate

$$\frac{\lambda_1 E_1 + \lambda_2 E_2 + \cdots + \lambda_k E_k + \lambda_{k+1} E_{k+1}}{1 + \lambda_1 h_1 + \lambda_2 h_2 + \cdots + \lambda_k h_k + \lambda_{k+1} h_{k+1}}.$$

Hint: Use **Fact 1** again.

b) Use **(a)** to show that the profitability of *any* item included in the diet is greater than the intake rate of the diet itself.

c) Explain why these observations make sense.

9. Suppose that a forager could select only one prey type for its diet. Assume that the profitability of item 1 is greater than that of item 2 (i.e., $E_1/h_1 > E_2/h_2$). Is it the case that the net caloric intake for a diet consisting of only the first item, $\lambda_1 E_1/(1 + \lambda_1 h_1)$, is greater than the net caloric intake rate for a diet consisting of only the second item, $\lambda_2 E_2/(1 + \lambda_2 h_2)$?

a) Using the information in **Exercise 7**, calculate the net caloric intake rate for a diet consisting of crabs *alone* and then calculate the rate for a diet of worms *alone*.

b) Show that the most profitable item does not produce the best single-item diet. Why does this happen?

c) However, even in this odd situation, verify that the diet obtained by using the optimal foraging rules has a higher net caloric intake rate than either single-item diet.

10. a) Assume that prey items are ranked by profitability. How large must the encounter rate λ_1 for the first prey item be to make the optimal diet consist only of the first ranked item? (Your answer should be given in terms of E_1, E_2, h_1, and h_2.)

b) Apply your result to the data in **Exercise 3c** to determine the minimum encounter rate λ_L for large snails so that small snails should be excluded from a lizard's optimal diet.

3.3 Habitat Quality and Breadth of Diet

While the optimal number of prey types will vary from predator to predator, the model also predicts that it will vary for the same animal in different habitats, as we shall see. Such predictions are crucial; they provide ways to test the accuracy of the model.

We can determine the general effect of habitat on breadth of diet using the following analysis. A poorer habitat is one in which the number of prey is smaller but other characteristics are unchanged. That is, profitability values and handling times are the same in both good and poor habitats, only the encounter rates, λ_i, change. How does this affect the net caloric intake rate, E/T?

Let E_1, h_1, and λ_1 represent the net energy value, the handling time, and the original encounter rate for the single prey type in the diet. Suppose that as the seasons change, the encounter rate λ_1 increases. Note that the profitability of the item remains the same, E_1/h_1. With more prey available, we expect that the net caloric intake rate will increase, that is, $\lambda_1 E_1/(1+\lambda_1 h_1)$ is an increasing function of λ_1. This is verified by taking the derivative with respect to λ_1, regarding E_1 and h_1 as constants. Check that

$$\frac{d}{d\lambda_1}\left(\frac{\lambda_1 E_1}{1+\lambda_1 h_1}\right) = \frac{E_1}{(1+\lambda_1 h_1)^2}.$$

Since this derivative is clearly positive, the intake rate is an increasing function of λ_1.

For multiple-item diets, the same is true. If any encounter rate λ_i of an item in the diet increases, then the intake rate does as well, as may be verified using partial derivatives. Check that

$$\frac{\partial}{\partial\lambda_i}\left(\frac{\lambda_1 E_1 + \cdots + \lambda_k E_k}{1+\lambda_1 h_1 + \cdots + \lambda_k h_k}\right) = \frac{E_i(1+\lambda_1 h_1 + \cdots + \lambda_k h_k) - h_i(\lambda_1 E_1 + \cdots + \lambda_k E_k)}{(1+\lambda_1 h_1 + \cdots + \lambda_k h_k)^2}.$$

This derivative is positive if and only if

$$E_i(1+\lambda_1 h_1 + \cdots + \lambda_k h_k) > h_i(\lambda_1 E_1 + \cdots + \lambda_k E_k),$$

which is equivalent to

$$\frac{E_i}{h_i} > \frac{\lambda_1 E_1 + \cdots + \lambda_k E_k}{1+\lambda_1 h_1 + \cdots + \lambda_k h_k},$$

which is true since the profitability of any item included in the diet is greater than the intake rate of the diet itself (see **Exercise 8**). Thus:

As any or all of the encounter rates increase, the net caloric intake rate for any diet increases, even though the profitability of the items has not changed.

Consider what this last statement means when applying the rules of optimal foraging. Another item is added to the diet only if its profitability exceeds the net caloric intake rate of the diet of higher-ranked items. If the intake rates of all diets increase because of increased encounter rates, then lower-ranked items are less likely to be included in an optimal diet. The profitability of lower-ranked items is less likely to exceed the increased intake rate of the diet of higher-ranked items.

This observation gives a way to test the optimal foraging model. The model predicts that if encounter rates are increased, an optimally foraging animal will be more selective and narrow the breadth of its diet. It can afford to be choosy in a plentiful environment. On the other hand, if encounter rates are decreased, then the diet breadth should widen as the forager becomes less selective. Beggars can't be choosers.

Exercises

11. Imagine a cardinal making foraging decisions in the summer and the fall. In the summer, there are many juicy caterpillars, plenty of berries, and a reasonable supply of seeds. By autumn, most of the caterpillars have turned into butterflies and departed, most of the berries have been eaten, and only the seeds remain available, as indicated in **Table 3** of hypothetical data.

Table 3.
The cardinal's diet in summer and autumn.

Item	E kcal	h min	E/h kcal/min	Rank	Summer λ #/min	Fall λ #/min
caterpillars	20	0.1			10	1
berries	15	0.1			30	2
seeds	20	0.5			10	9

a) Calculate the optimal diet for the plentiful summer season and its associated net caloric intake rate.

b) Calculate the optimal diet for the resource-poor autumn season and its net caloric intake rate.

c) How do the two diets differ in breadth and rate of intake?

12. a) Reconsider **A Diet for Blue Jays** (see **Table 1**), where an optimal diet was calculated to be worms and moths. How high would the encounter rate for worms have to be to exclude moths from the jay's diet?

b) Suppose that the encounter rate for worms were lowered to $\lambda = 0.1$ per minute. What is the greatest encounter rate for moths that ensures grubs would still be included in the jay's optimal diet?

13. Richardson and Verbeek [1986] studied the foraging behavior of Northwestern Crows (*Corvus caurinus*), which feed extensively on littleneck clams (*Venerupis japonica*). Crows expend a substantial amount of time and energy to dig up, open, and extract the edible parts of these clams. (After digging the clams out of the sand, the crows fly with them to rocky areas, where they drop the clams to crack them open.) Casual observation of shells in the drop area indicated that the crows fed preferentially on larger clams. Let M denote the dry mass (in mg) of the soft tissue of the clam and L the length (in mm) of the clam. Richardson and Verbeek estimated the mass as a function of length to be $M = 0.016L^{3.47}$. Each milligram of mass provides 2.0 joules of energy. Crows were estimated to have a 75% efficiency in energy assimilation. The energy expended by the crows on all aspects of handling a clam of length L was estimated to be $E_{h_L} = 0.936L^{1.58} + 362$ joules, with a corresponding handling time $h_L = 0.13L^{1.58} + 15.3$ seconds.

a) Find a general expression for the net energy content of a clam of length L mm.

b) Find a general expression for the profitability of a clam of length L mm.

c) Very few clams larger than 38 mm were available. Crows selected clams whose lengths were in the 20 to 38 mm range. Use calculus to determine the length in this range with the maximum profitability.

d) Suppose that, for convenience, the clams are sorted into the discrete size categories in **Table 4**. Fill in the net energy, handling time, and profitability for each category.

e) Using the given encounter rates, what size clams should be included in an optimal diet, and what is the corresponding net rate of energy intake?

Table 4.

Data for clams.

Size mm	Net Energy j	Handling Time s	Profitability j/s	Encounter Rate items/s
20				0.0093
23				0.0098
26				0.0111
29				0.0129
32				0.0107
35				0.0049
38				0.0004

14. Assume that you are a biologist. You have available a supply of caged cardinals and supplies of the caterpillars, berries, and seeds described in **Exercise 11**. You are free to place the birds in large experimental cages and to supply them with whatever type and abundance of food you want to. Design an experiment testing whether or not the cardinals forage according to the optimal foraging model outlined so far.

4. Testing the Model Using Prey Density

Conducting field or laboratory experiments to test optimal foraging theory is a complicated business, if for no other reason than there are so many variables involved. Each potential prey item has its own net caloric value E_i, handling time h_i, and encounter rate λ_i, which must be measured or estimated.

Researchers have been extremely clever in devising ways to make such estimates and in using controlled situations in a laboratory setting. We will use an experiment conducted by Werner and Hall [1974] to illustrate this process. They worked with bluegill sunfish, *Lepomis macrochirus*, which are known to select prey on the basis of size. These freshwater fish eat a wide range of prey, from small water snakes down to almost microscopic arthropods.

In an experimental pool, sunfish were allowed to select prey that were either large, medium, or small *Daphnia magna*. *Daphnia* are tiny swimming arthropods, commonly called water fleas. The size classes used in the experiment were biologically meaningful: *Daphnia* grow in discrete size intervals, or *instars*, because they molt. Werner and Hall were able to test optimal foraging theory in the following way.

Daphnia were sorted into size classes by washing them through screens with different size mesh. Populations in each size class were counted by sucking them up into a long thin glass tube.

The various size *Daphnia* were ranked by their profitability. This turns out to be easy, because handling time for all sizes of *Daphnia* was very similar, about 1.2 seconds per item. So h_i is actually a constant value, h, for all prey types. Because the caloric content of *Daphnia* is directly proportional to size, it follows that

$$\frac{E_L}{h} > \frac{E_M}{h} > \frac{E_S}{h},$$

where E_L, E_M, and E_S are the net caloric values of large, medium, and small *Daphnia*. That is, the larger the *Daphnia*, the greater its profitability. (Note the importance of constant handling time in this result and contrast it to **Figure 2**.)

Specific densities of each size class were added to a pool containing 10 hungry bluegills. The bluegills were allowed to feed for a predetermined amount of time. It was impossible to accurately observe which size classes were being eaten. So, at the end of each experiment, the bluegills were netted and sacrificed, and their stomach contents were examined for the number of *Daphnia* in each size class. Werner and Hall could now compare the actual diets of the fish with those predicted by optimal foraging theory.

In a series of trials, Werner and Hall manipulated the encounter rates of the prey items. They put the same number of *Daphnia* of each size class in the pool with the bluegills in either high or moderate densities. Because smaller *Daphnia* cannot be seen as far away as larger ones, the actual encounter rates were different for each size. In particular, if λ represented the encounter rate for the largest *Daphnia*, then Werner and Hall determined that the encounter rates for the medium and small *Daphnia* were 0.71λ and 0.27λ, respectively.

Optimal foraging theory predicts that varying λ will cause the sunfish to be more or less selective in their diet. The largest prey type should always be included in the bluegill's diet, since it is the most profitable. But the second-ranked prey type (medium) should not be included in the bluegill's diet if its profitability is less than the net caloric intake rate for a diet of the large prey.

That is, medium *Daphnia* should be disregarded if

$$\frac{\lambda E_L}{1+\lambda h} > \frac{E_M}{h}.$$

Solving for λ, the bluegill's diet should consist only of large *Daphnia* if

$$\lambda > \frac{E_M}{h(E_L - E_M)}.$$

Since h is known, and since the net caloric content is proportional to the size of the prey, Werner and Hall were able to determine the critical value of λ to be approximately 0.345 of the largest *Daphnia* per second. Above this value, the theory predicts that only large *Daphnia* will be selected.

Lowering the encounter rate λ by putting fewer prey in the feeding pool corresponds to decreasing the quality of the habitat. Optimal foraging theory predicts that in poor habitats, the bluegill will be less selective. They can increase the ratio E/T by lowering the search time component of T because search time is a significant factor here. In fact, the model predicts that all three prey sizes should be included in the bluegill's diet as long as the profitability of the smallest *Daphnia* is greater than the net caloric intake rate of a diet consisting of large and medium *Daphnia*, that is, whenever

$$\frac{\lambda E_L + 0.71\lambda E_M}{1+\lambda h + 0.71\lambda h} < \frac{E_S}{h}.$$

Solving for λ, the bluegills should be unselective and eat all prey encountered regardless of size if

$$\lambda < \frac{E_S}{h(E_L + 0.71E_M - 1.71E_S)}.$$

Werner and Hall determined this value of λ to be approximately 0.034 encounters per second. Encounter rates below this value should induce the bluegills to select all three size classes.

For values of λ between 0.034 and 0.345 encounters per second, the theory predicts that the large and medium *Daphnia* will be selected whenever encountered, while the small size will be ignored. Because the bluegill should eat all large and medium *Daphnia* encountered, the ratio of large to medium *Daphnia* actually eaten should be the same as the ratio of their encounter rates: 1 to 0.71.

Similarly, at the lowest encounter rates, $\lambda < 0.034$, because the bluegill should eat all three sizes of *Daphnia* as they are encountered, the ratios of *Daphnia* actually eaten should be the same as the ratios of their encounter rates: 1 to 0.71 for large to medium and 1 to 0.27 for large to small. However, in the low-density case, Werner and Hall chose to raise the encounter rates for the medium and small *Daphnia* so that all three types were encountered equally often. Under these circumstances, the model predicts that all three types would

Table 5.
Predicted and observed diet ratios for bluegill feeding on three sizes of *Daphnia*. Adapted from Werner and Hall [1974, Tables 5 and 6].

Density	Predicted Diet Ratios			Observed Diet Ratios		
	Large	Medium	Small	Large	Medium	Small
High	1.00	0.00	0.00	1.00	0.23	0.05
Intermediate	1.00	0.71	0.00	1.00	0.58	0.04
Low	1.00	1.00	1.00	1.00	0.90	0.90

be eaten equally often, that is, in a 1-to-1 ratio for all types. The predictions and the actual results obtained by Werner and Hall are provided in **Table 5**.

In each case, the predicted and actual diets are quite similar. The bluegills do change their foraging behavior in response to changing encounter rates. Though the selectivity is not precisely what optimal foraging theory predicts, it is clearly in the correct direction for what we expect from optimally foraging sunfish. At high and intermediate densities of prey, they select *Daphnia* very differently than they encounter them.

Exercises

15. In the low density experiment, Werner and Hall manipulated the encounter rates so that all prey sizes were encountered equally often. Using the data given in this section, if there were 20 large *Daphnia* in the pool, how many medium and small *Daphnia* were present?

16. Suppose that you were going to carry out experiments similar to the ones that Werner and Hall carried out with bluegills. You use four size classes of *Daphnia*, with class 1 being the largest and 4 the smallest. You determine that handling time is constant for all four size classes, $h = 1.0$ seconds per item. When equal numbers of each prey size are presented, you find that the smaller sizes are encountered less frequently because they cannot be seen as easily as the larger sizes. In particular, you find that if $\lambda_1 = \lambda$, then $\lambda_2 = 0.75\lambda$, $\lambda_3 = 0.50\lambda$, and $\lambda_4 = 0.25\lambda$. The net caloric content of the largest *Daphnia* is $E_1 = E$, then the net caloric content of the other sizes are $E_2 = 0.50E$, $E_3 = 0.40E$, and $E_4 = 0.25E$.

a) What is the smallest value of λ that leads to a prediction that the bluegills will select only the largest *Daphnia*?

b) What is the smallest value of λ that leads to a prediction that the bluegills will select only the largest two classes of *Daphnia*?

c) What is the smallest value of λ that leads to a prediction that the bluegills will select the largest three classes of *Daphnia*?

d) Suppose that you adjust the number of prey of sizes 2 through 4 so that they are encountered at the same rate as the largest size. Find that value of λ below which all four size classes will be eaten, that is, below which the bluegills are entirely unselective.

17. Sutherland [1982] conducted a study of the foraging behavior of oystercatchers (*Haematopus ostralegus*) feeding on various size cockles (*Cerastoderma edule*). The handling time h (measured in seconds as the time from when a bird found a cockle to when it resumed searching or changed its behavior) varied according to the length of the cockle, x, in mm. Sutherland estimated $h = 0.927x - 2.71$ seconds for cockles in the 20 to 40 mm range. Sutherland used the *ash-free dry weight* of various sized cockles to measure indirectly their net energy as a function of length. Ash-free dry weight is typically determined as follows. The dry weight would be determined first by drying each cockle in a low temperature oven until it had 0 moisture content. Next, the ash weight would be measured after each cockle was burnt in a furnace until only ash remained. Subtracting the ash weight from the dry weight produces the ash-free dry weight. Since the ash contains mostly non-organic components (i.e., minerals) that are not likely to contribute to the energy content of the organism, the ash-free dry weight is a measure of the energy content of the organism. There is no universal relationship between either ash-free dry weight or dry weight and calories, because each organism has a different composition of protein, fat, carbohydrates, and minerals producing different amounts of calories per gram. The assumption in the article is that the composition in each individual does not vary greatly within the same species. In this case, ash-free dry weight is an indirect measure of the energy contained in the cockle. Sutherland found that ash-free dry weight w in grams was given by $w = 0.000031x^{2.624}$, where x is length in mm. (Equations are adapted from [Sutherland 1982, Figure 1].)

a) Using w as an estimate of E, determine the profitability of a cockle as a function of its length.

b) Which size in the 20 to 40 mm range used by Sutherland is the most profitable? Hint: Use calculus to find the maximum profitability.

c) Sutherland showed that given the encounter rates of the various sizes of cockles available, oystercatchers had "a disproportionate number" of large cockles in their diet. Is this what you would expect given the calculations above? Does his observation tend to support optimal foraging theory?

5. Recognition Time

5.1 Taking the Time to Recognize Prey

Predators must not only "look towards" their potential prey but must also "recognize" them as prey. This is true whether or not a sense of vision is being used by the predator. For hearing, smell, or any other sensory system, there is a finite amount of time required to pick those cues indicating prey out from the background cues provided by the environment. Young deer (fawns) have a dappled pattern of light spots on a dark background, which makes them extremely difficult to see if they stand motionless in a light-dappled forest. Fishes may be transparent or coated with silvery scales that reflect the blue color of the water. A white rabbit may be very difficult to notice when it is sitting on a snowy field, but the same white rabbit would be obvious on an asphalt parking lot. It is the combination of the type of animal and its background that determines recognition time. Our eyes are also far better at spotting moving objects than at recognizing motionless ones. Prey therefore often "freeze" when a predator approaches, which increases the time it takes for a predator to recognize it. To avoid having to model the behaviors of both prey and predator, we will consider only situations in which prey cannot behave to reduce its recognition time considerably.

To understand the effect of recognition time on average net caloric intake rates, consider the following situation. Suppose that a predator is foraging at an optimal rate of E/T. What happens if additional prey types are introduced to the habitat? Some prey type, A, may have a higher profitability, E_A/h_A, than the current value of E/T. Another prey type, B, may have lower profitability, E_B/h_B, than E/T.

Assume first that *no recognition time* is required by the predator. From the rules for optimal foraging, it is clear that prey type B should be avoided; including it in the diet would lower the current value of E/T. (This, of course, is true even if recognition time is involved.) On the other hand, prey type A should be included in the diet, as it would increase the average net intake.

But what if *recognition is not immediate*? That is, what if total foraging time, T, includes not only search and handling times but also a component of recognition time? Then introducing the additional prey types A and B into the habitat requires the predator to spend additional time distinguishing between A which is to be included in the diet, B which is to be avoided, and all other types of objects in the environment. This additional recognition time has the effect of *lowering* the overall average net rate of intake, since it increases the T component of E/T. If the profitability of prey type A is not much larger than the old optimal average rate of intake, E/T, then the increase in recognition time (and hence in T) may even outweigh any gain in increased average profitability obtained by including A in the diet. Paradoxically, the new optimum rate, E/T, may be lower than the old optimum, despite the habitat being "enriched" by prey type A.

5.2 Including Recognition Time in the Model

By modifying the basic model equations slightly, it is possible to account for the effect of recognition time in calculating caloric intake rates and optimal diets. Just as handling times and encounter rates vary for each type of prey, we now assume that the time to recognize a prey item will also vary with the type. Let r_i denote the recognition time for items of prey type i.

Next, assume, as in **Section 3.1**, that t minutes are devoted to searching for a diet of the k most profitable items. If λ_i is the encounter rate for the ith item in this diet, then in t minutes of searching, an average of $\lambda_i t$ items of type i will be encountered. Each of the $\lambda_i t$ items requires an expenditure of r_i minutes of recognition time for a total of $r_i \lambda_i t$ minutes of recognition time spent for this type. Hence, the total recognition time required for items of all k types in the diet in t minutes of searching is

$$r = \lambda_1 r_1 t + \lambda_2 r_2 t + \cdots + \lambda_k r_k t = t(\lambda_1 r_1 + \lambda_2 r_2 + \cdots + \lambda_k r_k).$$

Total time spent foraging will now be denoted by $\mathcal{T}$, and it includes a component of recognition time as well as search and handling times. Using the expressions developed for handling time (review **Section 3.1**), we have

$$\begin{aligned} \mathcal{T} &= t + h + r \\ &= t + t(\lambda_1 h_1 + \lambda_2 h_2 + \cdots + \lambda_k h_k) + t(\lambda_1 r_1 + \lambda_2 r_2 + \cdots + \lambda_k r_k) \\ &= t(1 + (\lambda_1 h_1 + \lambda_2 h_2 + \cdots + \lambda_k h_k) + (\lambda_1 r_1 + \lambda_2 r_2 + \cdots + \lambda_k r_k)) \\ &= t(1 + \lambda_1(h_1 + r_1) + \lambda_2(h_2 + r_2) + \cdots + \lambda_k(h_k + r_k)). \end{aligned}$$

Consequently, the net caloric intake rate of a k item diet is

$$\begin{aligned} \frac{E}{\mathcal{T}} &= \frac{t(\lambda_1 E_1 + \lambda_2 E_2 + \cdots + \lambda_k E_k)}{t(1 + \lambda_1(h_1 + r_1) + \lambda_2(h_2 + r_2) + \cdots + \lambda_k(h_k + r_k))} \\ &= \frac{\lambda_1 E_1 + \lambda_2 E_2 + \cdots + \lambda_k E_k}{1 + \lambda_1(h_1 + r_1) + \lambda_2(h_2 + r_2) + \cdots + \lambda_k(h_k + r_k)}. \end{aligned}$$

When should an additional item be included in the diet? The same analysis used in **Section 3.2** shows that prey type $k+1$ should be included in the diet if and only if

$$\frac{\lambda_1 E_1 + \lambda_2 E_2 + \cdots + \lambda_k E_k}{1 + \lambda_1(h_1 + r_1) + \lambda_2(h_2 + r_2) + \cdots + \lambda_k(h_k + r_k)} < \frac{E_{k+1}}{h_{k+1} + r_{k+1}}.$$

Call the ratio $E_i/(h_i + r_i)$ the *generalized profitability* and denote it by GP. Then this last inequality says that prey type $k+1$ should be included in the diet if and only if the net caloric intake rate $E/\mathcal{T}$ for the diet of the first k items is less than the generalized profitability GP for the $(k+1)$st item. This is the analog of Rule 2 for optimal foraging developed in **Section 3.2**. The other rules are similarly modified.

5.3 Distinguishing Monarch and Viceroy Butterflies

Toxic or unpalatable animals often possess bright warning coloration to warn off potential predators or to remind those who have earlier tasted that type of toxic animal of the likely repercussions. Sometimes, over evolutionary time, palatable animals will come to resemble toxic animals. Such palatable animals may gain some protection from predators in this fashion (they are called *Batesian mimics*). For example, monarch butterflies are toxic and induce vomiting when consumed by blue jays. Viceroy butterflies are harmless Batesian mimics of monarchs. Viceroys are fine blue-jay food.

Example: A Blue Jay's Diet Reconsidered. Recall that in the earlier example based on the data in **Table 1**, optimal foraging theory predicted that a blue jay's diet should consist of worms and moths and would have a net caloric intake rate of 29.7 kcal/min. In this example, suppose that we enrich the jay's hypothetical environment by adding two types of butterflies: the viceroy and monarch. (See **Table 6**.) The viceroy has a high profitability, $P = E/h = 30/0.6 = 50$ kcal/min. The toxic monarch has negative profitability since the jay will regurgitate it when eaten.

Table 6.
An expanded list of possible items in the jay's diet. Compare with **Table 1**.

Item	E kcal	h min	r min	λ #/min	P kcal/min	GP kcal/min
worm	162	3.6	0.0	0.2	45.0	45.0
moth	24	0.6	0.3	3.0	40.0	26.7
grub	40	1.6	0.0	3.0	25.0	25.0
viceroy	30	0.6	2.4	1.0	50.0	10.0
monarch	−24	0.6	2.4	1.0	−40.0	−8.0

Notice that it requires a substantial amount of recognition time to distinguish the two butterflies. Further, the jay must now spend some time making sure that the moth is not a monarch. This recognition time changes the value of these food items for the predator, as can be seen by comparing the profitability, P, with the generalized profitability, GP, in the last two columns of **Table 6**. What is the optimal diet, taking recognition time into account?

Solution. Worms have the highest generalized profitability. The net intake rate for a diet of worms only is

$$\frac{E}{T} = \frac{0.2(162)}{1 + 0.2(3.6 + 0.0)} \approx 18.8 \text{ kcal/min}.$$

The generalized profitability of moths is 26.7 kcal/min and is greater than the net intake rate of the diet of worms. So moths are included in the diet. Notice how recognition time now enters the calculation for the net intake rate for a diet of worms and moths:

$$\frac{E}{T} = \frac{0.2(162) + 3.0(24)}{1 + 0.2(3.6 + 0.0) + 3.0(0.6 + 0.3)} \approx 23.6 \text{ kcal/min}.$$

The generalized profitability of grubs (25 kcal/min) is higher than the intake rate from the diet of worms and moths, so grubs are now included in the diet, unlike in the earlier example. The net intake rate of this diet is

$$\frac{E}{T} = \frac{0.2(162) + 3.0(24) + 3.0(40)}{1 + 0.2(3.6 + 0.0) + 3.0(0.6 + 0.3) + 3.0(1.6)} \approx 24.3 \text{ kcal/min.}$$

Neither the viceroy nor the monarch should be included in the optimal diet, since their generalized profitabilities fall below 24.2 kcal/min. Notice that the optimal diet in the "enriched" habitat has a lower net caloric intake rate than the 29.7 kcal/min rate when there were no butterflies present.

Exercises

18. Consider the hypothetical data of **Table 7** for potential diet items for a gull.

Table 7.
Hypothetical data for **Exercise 18**.

Prey	E kcal	h min	r min	λ #/min	P kcal/min	GP kcal/min
shore worms	27	0.7	0.2	2.0		
crabs	160	3.9	0.1	0.5		
mussels	46	1.7	0.2	6.0		
small snails	50	1.7	0.3	10.0		
large snails	150	5.0	0.3	6.0		

a) Fill in the profitability, P, and the generalized profitability, GP.

b) Find the optimal diet and the corresponding net caloric intake rate *without* taking recognition time into account.

c) Now find the optimal diet and the corresponding net intake rate accounting for recognition time.

d) Compare and explain your results.

19. Return to **Table 6** and find the optimal diet for the jay ignoring recognition time. Has the addition of the butterflies "enriched" the habitat from this point of view, when compared to the results of the original example? Also compare this to the result that takes recognition time into account.

6. Modeling Food Patches

6.1 Different Patch Types

The natural environment, with the exception of vast expanses of desert or prairie or ocean, is a patchy entity. We rarely see a uniform distribution of any plant or animal. Specific plants grow in spots where the soil moisture, soil chemistry, acidity, and supply of sunlight all favor growth of that particular organism. Patches of appropriate conditions may be very small but commonly will support a number of plants of the same type. Herbivorous animals make their living by eating plants, and so herbivores occur in patches just as their food plants do. Carnivorous animals survive by eating herbivorous animals, and so they, too, will be in patches wherever their prey are located.

Actively searching predators normally move from patch to patch as they look for food. Just as individual prey items differ from one another in their profitability, habitat patches in which food may occur differ from one another in size and quality. An optimally foraging predator should forage preferentially in those patches with the greatest potential profitability, and only search in patches of poorer quality when the more valuable patches are present in a low density.

Just as individual prey items can be characterized in terms of their quality, various types of patches can be characterized according to profitability. We can then ask the following:

- How many different types of patches of food should a predator visit out of any given array of patches?
- How long should a predator spend searching within patches of each type?

There are three different types of patches that must be distinguished before attempting to answer these questions.

1. Some patches retain a constant potential profitability regardless of the presence of a predator. This is generally uncommon in nature, because a predator that removes prey from a patch normally decreases its future profitability. But an approximate example might be the enormous population densities of lemmings in the Arctic at certain times in their multiyear predator/prey oscillation cycle, which produce such large groups of these creatures that they undertake their famous migrations culminating in mistaking the ocean floor for a lake. At such times, the lemmings may be an essentially infinite food patch for arctic owls [Pitelka et al. 1955].

2. Some patches are depleted by the presence of a predator feeding in them. The longer the predator is there removing prey items, the less will be the future supply of prey in the patch. Since profitability declines over time, a predator must decide when to leave the patch. For example, worker honey bees each tend to specialize on particular species of flowers from which they collect nectar, and floral patches can be rapidly depleted of their daily nectar supply, so that patches become economically valueless until the following day [Kolmes 1990].

3. Some patches might change in their profitability regardless of the presence of a predator. If, for instance, the time of day or temperature of the air were closely related to the availability of prey, a predator might have to leave a

patch based upon changes in profitability unrelated to its own activity. For example, northwestern crows apparently optimize the size of the littleneck clams that they collect, but they cease foraging when the incoming tide deeply submerges an area of clams [Richardson and Verbeek 1986].

6.2 Optimal Foraging in Patches

We now examine what optimal foraging strategy has to say about Cases 1 and 2 in the previous subsection. Case 3 becomes quite complicated and little detailed analysis is possible at this level.

In Case 1, the quality of the patches remains constant. The optimal foraging strategy is obvious: Determine which patch is highest in quality and stay there. Although this situation seldom arises in nature, it can occur in the laboratory. Pigeons in Skinner boxes peck at keys associated with functionally infinite bins of food pellets. The pigeons are rewarded with food at rates that are determined by which key they peck in the box. The pigeons quickly learn to press only the key that yields the greatest amount of food (akin to remaining in the patch of highest quality). That is, the pigeons forage optimally.

Case 2 is more interesting. Here the assumption is that the predator depletes the patch as it continues to forage. As prey are consumed, there are fewer left to encounter in the patch. Consequently, the predator encounters prey within the patch at an ever decreasing rate. If we graph $E(t)$, the net caloric intake, versus time (including the travel time to reach the patch), we expect rapid gains upon first entering the patch, followed by slow growth as a predator learns where it can feed within a patch, and eventual decline as more and more energy is expended searching for fewer and fewer resources (**Figure 3**). Also notice that as the predator travels to the patch, it experiences a net loss of energy.

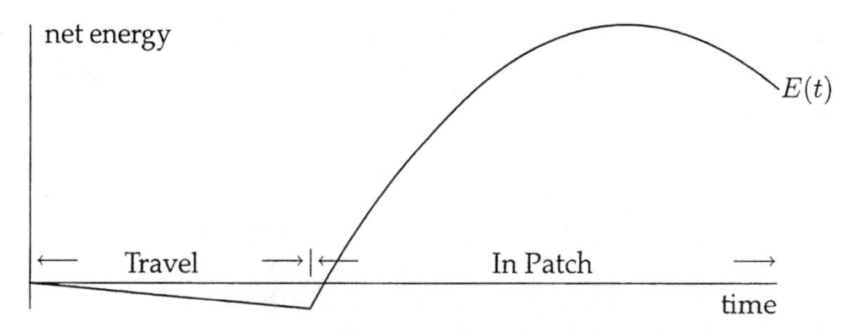

Figure 3. Cumulative net energy gain vs. time, in a single patch.

At what point should a predator move to another patch? Certainly it must leave before it begins to lose energy searching for the very sparsely distributed prey remaining in the patch. We might be tempted to conclude that the predator should stay in the patch until the maximum net gain in energy (highest point on the curve) has been reached within that patch. But most animals leave much sooner than this. What do they know that we don't? What have we failed to account for?

In the discussion so far, we have concentrated on the cumulative net gain within a patch. Of course, the optimal forager is trying to maximize net gain over all patches in the habitat. More precisely, the forager seeks to maximize the *average net caloric intake rate*. This rate is just the quantity E/T again, where total foraging time T now *includes the travel time between patches* as well as the time spent within the patch. Of course, the average rate of net gain varies with the amount of time spent within the patch. There is a simple graphical way to interpret this average rate of change.

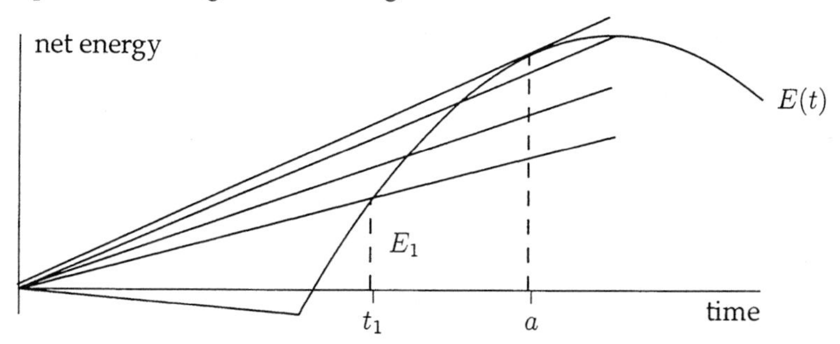

Figure 4. The average net caloric intake rate, E/T, at the time t_1 in the patch is the net energy E_1 divided by the time t_1. This is the slope of the line from the origin to the point (t_1, E_1) on the curve. The maximum intake rate occurs when the slope is the steepest: at time $t = a$ where the tangent line passes through the origin.

The average energy intake rate for a particular patch at any time t_1 is just the cumulative energy gain, E_1, up to that moment, divided by the time t_1. But this is just the slope of the line through the origin and the point at t_1 on the net energy graph. The optimal forager is presumed to maximize E/T, the average rate of intake. Mathematically, this corresponds to finding the maximum slope of all lines that pass through the origin and a point on the cumulative energy graph. **Figure 4** provides evidence for the following observations:

1. E/T is maximized at the unique point on the curve where the tangent line passes through the origin. (This is the so-called **marginal value theorem**. It was employed in this context for the first time by Charnov [1976], who adapted the result from economics.)

2. The time that maximizes E/T occurs before the cumulative net energy graph reaches its maximum.

These two observations can be proven using calculus. To do so, we must assume that:

- $E = E(T)$ is a twice differentiable function of T, at least for the time interval in the patch.

- The earlier statement, "$E(T)$ rises quickly after entering the patch and then more slowly until it reaches a maximum and then begins to decline," means that $E(T)$ has a local maximum at some time $T = d$, so that $E'(T) > 0$ before d and $E'(T) < 0$ after d, so that $E(T)$ is concave down, $E''(T) < 0$.

Using these assumptions, we will prove that:

1. there is only one time $T = x$ when the tangent line to $E(T)$ passes through the origin, and

2. at this same moment the average net energy intake rate is a global maximum.

To prove (1), recall that because the tangent slope is the derivative, the equation of the tangent at $(a, E(a))$ is

$$y - E(a) = E'(a)(T - a).$$

Equivalently,

$$y = TE'(a) + E(a) - aE'(a).$$

The y-intercept of this particular line is $b = E(a) - aE'(a)$. So at a general point $(T, E(T))$, the y-intercept of the tangent line is the function,

$$b(T) = E(T) - TE'(T).$$

By assumption, $E(T)$ is twice differentiable, so $b(T)$ is continuous and differentiable.

Examine **Figure 5** to see why there is exactly one time, $T = x$, when the tangent line has intercept 0, that is, when $b(x) = 0$. Let $T = c$ be the time at which the animal first enters the patch. Then the intercept of the tangent line to the net energy curve, $b(c) = E(c) - cE'(c)$, is negative because $E(c)$ is negative and $E'(c)$ is positive by assumption. Later, when the E reaches its maximum at $T = d$, the intercept, $b(d) = E(d) - dE'(d)$, is necessarily positive because $E(d)$ is (else the animal would starve) and $E'(d) = 0$. Because $b(T)$ is continuous, the Intermediate Value Theorem guarantees that there is a time $T = x$ between c and d when $b(x) = 0$, that is, the intercept must be 0 at $T = x$.

Next, notice that $b(T) = E(T) - TE'(T)$ is a strictly increasing function because

$$b'(T) = E'(T) - E'(T) - TE''(T) = -TE''(T) > 0,$$

where the final inequality follows from the assumption that $E''(T) < 0$. But if $b(T)$ is strictly increasing, then $b(T) = 0$ *at most once.* (See **Exercise 20** if you are not familiar with this fact.) Since $b(x) = 0$, then $(x, E(x))$ is the *only* point on the curve where the tangent passes through the origin.

To prove (2), we will show that x is the only critical value of $E(T)/T$. We have

$$\frac{d}{dT}\left(\frac{E(T)}{T}\right) = \frac{TE'(T) - E(T)}{T^2} = \frac{-b(T)}{T^2}.$$

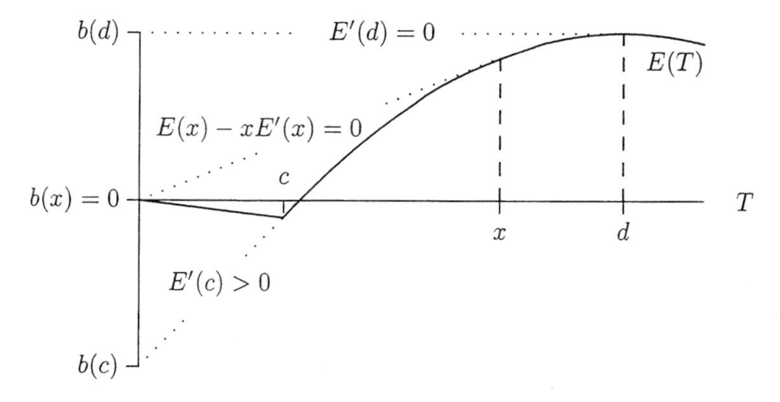

Figure 5. The intercept $b(T)$ of the tangent line to the net energy curve changes from negative to positive during patch residence time, so by continuity it must be 0 at some time $T = x$.

Since critical values of differentiable functions occur where the derivative is 0, we need $b(T) = 0$. But we just proved that $T = x$ is the only time when this happens. Further, we know from the proof of (1) that $b(T)$ is negative before $T = x$ and positive after. Thus, the derivative $-b(T)/T^2$ changes from positive to negative at x. So the first derivative test implies that $E(T)/T$ has a local maximum at $T = x$. Since $E(T)/T$ has only one critical point, its local maximum is a global maximum.

In summary:

The marginal value theorem shows that a predator should remain within each patch until its instantaneous rate of intake drops to the level of the average rate of intake.

Exercise

20. Prove the statement made in the argument above: If the differentiable function $b(t)$ is strictly increasing (i.e., $b'(t) > 0$), then $b(t) = 0$ for at most one value of t. Hint: Assume that this is not true, that is, that there exist values t_1 and t_2 with $t_1 < t_2$ such that $b(t_1) = b(t_2) = 0$. Use Rolle's theorem or the mean value theorem to derive a contradiction.

6.3 Testing the Patch Model: Varying Travel Time

By varying particular aspects of a patchy habitat, it is possible to test the model by making predictions about the results of these changes. In particular, how should an optimal forager behave if the travel time between patches is decreased (increased)? This corresponds to increasing (decreasing) the value of each patch. What do you know about selectivity in such cases? The answer is graphically clear in **Figure 6**.

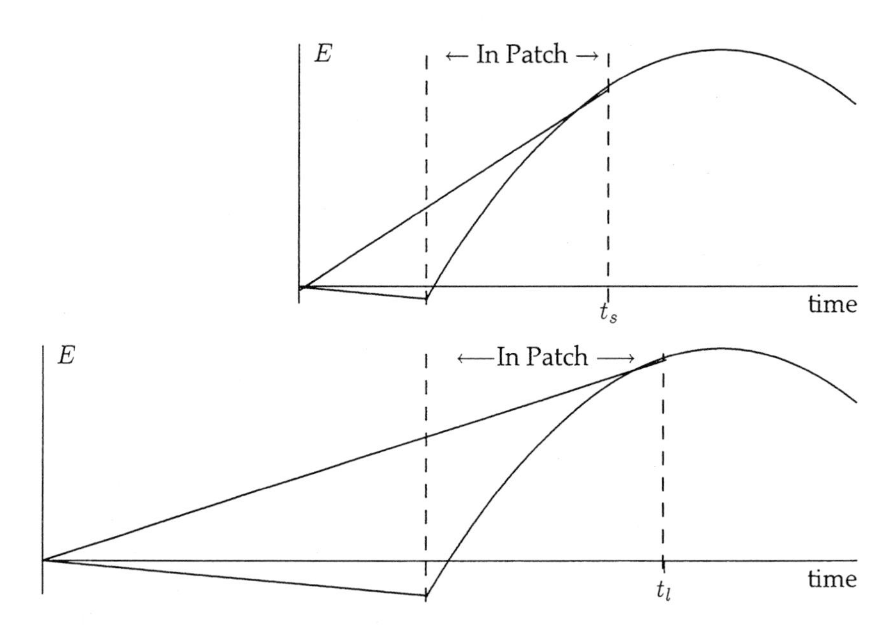

Figure 6. To determine the effect of travel time on average net caloric intake rate, use identical energy-intake curves in two patches. Align the time of entrance into both the patches. Habitats with shorter travel times between patches (upper figure) will result in optimal residence times that are shorter.

When travel times are shortened, the optimal time spent in the patch is also shortened (to t_s); the predator has become more selective in the "better" habitat. The result is an increase in the average net intake rate (steeper slope of the tangent line). Conversely, when patches are widely separated, the optimal time in the patch is lengthened (to t_l) because it is quite costly to move to the next patch. The predator has become less selective. The result is a decrease in the average intake rate. Thus, we predict that time in patch varies directly with the travel time (or distance) between the patches.

Example: Cowie's Birds. In a laboratory situation, Cowie [1977] studied how small European birds called great tits moved from one food patch to another. His food patches were plastic cups filled with sawdust, with insect larvae (mealworms) that the birds happily consume. There were cardboard lids over the plastic cups, a loosely fitting cardboard lid on a cup corresponded to a patch of food that was easily entered (i.e., "short travel time between the food patches"). A tightly fitting cardboard lid took the birds far longer to pry off and corresponded to a food patch that was difficult to get to (i.e., "long travel time between patches").

Optimal patch choice predicts that as travel time between patches increases, a bird should spend longer within each patch exploiting its resources before moving on. Cowie found that although residence time in each "patch" (time at each food cup once the lid was off) did increase as the time expenditure needed to get into each patch increased, that the great tits were nonetheless spending longer in each patch than predicted. When Cowie calculated the energy involved in searching for food rather than the time involved (using metabolic rate data, costs of moving about, etc.), he got a better fit between predictions and data. In this experiment "commuting time" turned out to be a good predictor of bird behavior, but "commuting energy" was a better predictor.

Exercise

21. Suppose that a laboratory experiment were conducted to test the effect of varying the net energy so that the travel time between patches of food items was held constant. For the control group a fixed number of food pellets are placed in each patch. For the experimental group, the same number of pellets are used but their size is doubled. Thus, if the net intake function was $E(t)$ for the control group, then it was $2E(t)$ for the experimental group.

a) Use the model to predict how the time spent in the patches should change for the experimental group. Hint: If $T = a$ is the critical value of $E(T)/T$, then what is the critical value of $2E(T)/T$?

b) Explain why your answer makes biological sense!

7. Central Place Foraging

7.1 Round-Trip Costs of Provisioning Young

In the preceding sections we considered individual foragers feeding themselves. In such situations the forager must decide where to search for food, what foods to eat, and when to leave a patch. However, many animals must feed not only themselves but also their young. Parents must provision the young by returning with excess food. In the original model, the travel from a nest or perch to a patch was a one-way energetic cost. In the case of a parent provisioning its young, the energetic cost is for a round trip between the nest and the patch. This modification of the model is often referred to as *central place foraging* (CPF), because the animal must return to a central place after each foraging bout.

Obviously, round-trip costs will be greater than one-way costs, and the distance to the feeding area will determine, in part, how much food the parent must carry back to the offspring to make the round trip profitable. We can analyze this situation in the following way. As usual, E represents net energy. In this context, $E = E_l - E_h - E_r$, that is, the energy content of the food load E_l minus the energy expended by the parent in handling it E_h and minus the energy expended on the round trip E_r. Total time $T = t + h + r$ is the sum of search time s, handling time h, and round-trip traveling time r. Thus, the net energy intake rate is

$$\frac{E}{T} = \frac{E_l - E_h - E_r}{t + h + r}.$$

Notice how the distance r affects the model. The greater the distance of the item from the nest, the smaller E is and the larger T is. Thus, any load of food becomes less profitable the farther away it is from the nest. This allows us to test the model by making predictions about the food selection of adults provisioning their young. The smallest food loads contain very few calories, because E_l and hence E are small. Such loads are not profitable unless they are very close to the nest. On the other hand, larger loads contain more calories (E_l and hence E are large) and are profitable at greater distances from the nest. Thus, we expect that the average size of food loads selected by a provisioning parent to increase with the distance from the nest.

However, there are physical constraints with respect to the load size that can be carried by any parent. Birds, for example, must carry food to the young in their beaks or crops. African hunting dogs carry meat to their young in their stomachs and regurgitate the partially digested meat at the den.

In a test of the CPF model, Bryant and Turner [1982] calculated the optimal load sizes of house martins foraging for their offspring (see **Table 8**). Then they measured the actual loads fed to each chick by the returning parents and compared them to the predicted load sizes. The data agree with the predictions of CPF theory in a qualitative way: Mean load size increased with distance from the nest. But the observed load sizes did not agree with the values predicted by the theory, perhaps due to the difficulty in collecting accurate data in the field.

Table 8.
Data collected by Bryant and Turner [1982] for foraging house martins.

Foraging Distance (km)	Travel Time (min)	Optimal Load (mg)	Observed Load (mg)
0.10	0.32	32	52.6
0.45	1.43	43	55.2
1.00	3.18	52	65.6

7.2 Variations of the CPF Model

So far we have been concerned only with provisioning the young, but CPF models also apply to situations where that is not the immediate concern. Consider situations in which other items are returned from a patch of resources to a central place, or else situations where food is brought back but not fed to offspring, as is done in nest building. Another case might be where food is returned to a central place for reasons other than feeding offspring, including situations where food is stored or cached for later use. Squirrels are notorious for their ability to cache large numbers of seeds and nuts for use over the long cold winter. Other examples might include:

- bees foraging for pollen, which is returned to the hive;
- leaf cutter ants foraging for leaves that are brought back to the underground nest for use in cultivating a fungus; or
- kangaroo rats of the southwestern United States, which collect and store seeds in external cheek pouches for the trip back to the burrow.

Example: Chipmunk Load Size. The eastern chipmunk (*Tamias striatus*) gathers seeds, transports them in its cheek pouches, and caches them in its burrow for use during times of food shortage. By measuring the number of seeds removed at three distances from the cache site, Giraldeau et al. [1994] demonstrated that chipmunks collected larger loads as the distance from the burrow increased (see **Table 9**).

Table 9.
Results of data collected in the field for eastern chipmunks foraging for sunflower seeds at artificial feeding stations at three distances from the burrow.
Data extracted from Giraldeau et al. [1994, Figure 4].

Distance (m)	Mean load size without competitors (# of seeds)	Mean load size with competitors (# of seeds)
1–3	19	16
13–30	23	20
35–68	26	22

Notice that the mean load size decreased when potential competitors were present within the territory of the chipmunk, but the overall pattern of increased load size with increased distance remained the same. The change in load size in the presence of competitors points out that the simple models that have been constructed do not take all factors that affect foraging into account and indicate why field results may agree qualitatively, even if not precisely, with model predictions.

Exercises

22. In a study of food-hoarding in grey jays (*Perisoreus canadensis*), Waite and Ydenberg [1996] evaluated the rate maximizing strategy of jays transporting raisins to a cache site. Instead of varying the distance to the cache site, the researchers forced the bird to wait at a feeding platform for varying amounts of time before each additional raisin was presented to the bird.

a) How would you modify the expression for the CPF net energy intake rate, E/T, to account for the waiting time, w, and the energetic costs of waiting, E_w?

Table 10.
Data for grey jays hoarding raisins at field sites in Algonquin Provincial Park, Ontario, and Koyukuk, Alaska. Adapted from Waite and Ydenberg [1996, Figure 4].

Waiting time (s)	Algonquin Park Mean load size (# raisins)	Koyukuk, Alaska Mean load size (# raisins)
0	4.0	3.2
15	5.4	4.6
30	6.0	5.5
45	6.7	6.2

b) Based on your new equation, predict how load size would change for grey jays as waiting time increases.

c) Are your predictions supported by the data collected by Waite and Ydenberg (see **Table 10**).

23. Project Outline. Although many animals can be used to study CPF, in practice some species are considerably more difficult to observe than others. Obviously, it would be difficult to measure the distance to a feeding patch for a seabird that forages far out to sea. Likewise, it may not be practical to determine the load size of small insects carried in the beak of a small swallow. The choice of organism and the design of the study are often constrained by practical considerations. Squirrels or chipmunks can make good subjects. In situations where class time is limited, beavers (*Castor canadensis*) may make good subjects for optimal foraging studies because they leave direct evidence of their food choices (or building material choices) behind in the form of felled trees and stumps.

a) Select a site where recent beaver cuttings are evident. With a measuring tape or rope, mark off a transect beginning at the water's edge and extending perpendicular to the water for 30 meters or more into the forest. Determine which trees and stumps lie within one meter on either side of the tape. You now have a transect two meters wide and 30 or more meters long.

b) Within the transect, record the species of each tree or stump, its diameter (or circumference) at one foot above the ground (typical height beavers will begin their cuts), and the tree's distance from the water. Remember you must record both cut trees (stumps) as well as uncut trees. Why is this necessary?

c) Plot your data for cut trees with distance from the water on the horizontal axis and tree diameter (or circumference) on the vertical axis. Using a different symbol, plot your data for uncut trees on the same axes. Hint: To see an overall pattern, instead of plotting every point, subdivide the data into groups of trees using 2 to 5 meter intervals from the shore. Then plot the mean or median diameter for all the trees within each distance subinterval on your graph.

d) Does the distance from the water influence the size of the trees that the beavers cut? How does the size of the cut trees compare to the uncut trees? What other factors might play an important role in the choice of tree felled as distance from the shoreline increases?

e) Assuming you were able to identify tree species, what does optimal foraging theory predict about the beavers' preference for some types of trees? Is this supported by your data?

f) Are there other factors that might affect the choice of tree species felled?

24. Consider the three data sets in **Table 11**. Each shows the mean diameter of trees cut by beaver at certain distances from the shore.

Table 11.

Tree diameter versus distance from shore for (1) oak and (2) maple trees cut by beavers at Blue Heron Cove at the Quabbin Reservoir, MA [Jenkins 1980]. (3) Tree diameter versus distance from shore for trees cut by beavers at the Hanley Preserve, Seneca County, NY (Hobart and William Smith Student Data 1996).

1. Quabbin Reservoir, MA: Oaks										
Distance (m)	6	16	36	45	55	66	75	85	95	105
Diameter (cm)	31	20	25	14	9	7	8	6	7	8
2. Quabbin Reservoir, MA: Maples										
Distance (m)	5	15	25	35	45	55	65	75	85	95
Diameter (cm)	11	8	7	5	5	5	4	5	5	4
3. Hanley Preserve, NY										
Distance (m)	2	6	10	14	18	22	26			
Diameter (cm)	5	6	5	3	4	5	3			

a) Do these data show a consistent relationship between distance from shore and diameter of tree cut? Explain.

b) Do any of these data sets support the CPF model predictions that larger trees would be felled further from the shore? Explain.

8. Complications, Critiques, and Criticisms

8.1 The Problem of Nutrients

So far, we have acted as though prey were largely equivalent except for size. But every animal has a set of nutrient needs that go beyond just energy intake. A diet of potato chips and soda is adequate in calories of energy supplied, but it is unhealthy for a human because of the protein, vitamin, and mineral needs of our bodies. Vitamins are molecules that animals cannot synthesize within their own bodies but which are crucial to the proper functioning of the chemical reactions that support life. Minerals serve a variety of functions. A single mineral like calcium is required for bone and tooth growth, nerve impulse conduction, and muscle contraction. Therefore, every animal (humans being neither especially complicated nor simple nutritionally) must find an array of food items that will provide it with all of its noncaloric nutritional needs as well as meeting its daily calorie requirement.

If other nutrients are essential to a predator, then we expect to see in some cases the inclusion of prey with low energy profitability in the diet of predators. In some cases we may see the inclusion of such prey to the complete exclusion of some highly profitable prey.

Example: Redshank Diets. This seems to have been the case in an experiment [Goss-Custard 1977] on redshank (shore-birds). First, the redshank were presented solely with a diet of different size worms. As in the bluegill experiment, when densities of all sizes of worms were high, the redshank were seen to be optimal foragers and selected only the largest worms for their diet. As densities were lowered, the less profitable smaller worms began to be included in the diet, again confirming the optimal foraging model.

However, redshank preferred the smaller, less profitable *Corophium* (a small invertebrate with many jointed legs and antennae) to all sizes of worms. Since *Corophium* supply less energy per unit handling time than worms do, this appears to **contradict** the optimal foraging model. Perhaps the *Corophium* contain some essential nutrient for the redshank? Optimal foraging models have been developed to handle such situations. For a nice introduction to such attempts, see Stephens and Krebs [1986, 63–65].

8.2 Other Considerations

There are many other aspects of energy balance, movement, and nutrition that undoubtedly impinge on how various organisms have evolved to feed. We briefly describe a few such considerations below. The process of modeling optimal foraging behavior is far from complete.

Different types of foraging behavior incur different associated energetic costs. Consider for one example a hummingbird. These birds feed by sucking nectar up out of flowers to fill their crop. We might guess that a hummingbird would minimize the cost of traveling between flowers by staying at any one good flower that it found and completely filling its crop with nectar. However, although this minimizes "commuting distance," a full crop is a heavy load for a small hummingbird to carry. The energetic costs of remaining airborne increase rapidly as more nectar gets placed in a hummingbird's crop. Hummingbirds should therefore only fill their crops until the additional flight cost due to the weight load begins to equal the energetic value of the nectar sucked out of the flower, and should then stop even if their crops are not wholly filled

[De Benedictis et al. 1978].

A similar example is a shorebird called the bar-tailed godwit [Evans 1976]. These birds walk along ocean tidal zones and search for worms that live in L-shaped burrows in the sand. (The worms are *Arenicola marina*, commonly called lug worms.) When the worms are plentiful, the godwits walk along quite quickly. This brisk foraging walk expends a considerable amount of energy, but the energetic output is repaid by the large number of worms captured. When available prey become less dense, the godwits walk more slowly as they search, expending less energy.

High nutrient content may be a factor in predators selecting prey items that have low profitability. But sometimes predators seem to ignore the most profitable prey sizes of a single food type, as in one study of golden plovers [Thompson and Barnard 1984]. Though the largest worms were the most profitable, plovers often selected intermediate or smaller ones to eat. One possible reason for this behavior was that gulls regularly stole worms from the plovers. Observations indicated that plovers were most likely to lose the largest size worms to the gulls. In fact, over 60% of the largest size class was stolen when gulls were present. Interestingly, plovers would sometimes discard worms they had already collected, whether or not they were about to be attacked by gulls. The worms most often discarded were the largest, most profitable ones! This suggests that plovers modified their foraging behavior in response to the gulls.

Some animals hunt in groups or packs. Under what circumstances does group foraging benefit animals when compared to individual foraging? Is there an optimal pack size for such predators?

Example: Wild Dog Pack Size. The hunting behavior of African wild dogs (*Lyacon pictus*) was studied by Creel and Creel [1995]. The dogs most often hunted in a pack size of 10 adults. Creel and Creel plotted kilograms of prey killed per day per dog (which plays the role of E/T for the single foragers considered earlier) against pack size and found that a parabolic curve fit the data reasonably well (**Figure 7a**). The preferred pack size of 10 is near the *minimum* of this curve; the dogs seem to hunt in packs that minimize foraging success. The graph indicates that the dogs should hunt in packs as large as possible to maximize kg/dog/day.

However, the quantity kg/dog/day does not account for the dogs' foraging effort, which can be considerable. During a full-speed chase of prey such as impala, dogs may run at 40 to 60 km/h. If the quantity kg killed per dog per day per km chased is plotted against pack size, the data are reasonably approximated by a parabola with its peak at a pack size of about 12 (**Figure 7b**). This is much closer to the observed preferred pack size of 10. Other factors, such as defense of territory, clearly affect optimal pack size, but foraging success appears to play a role.

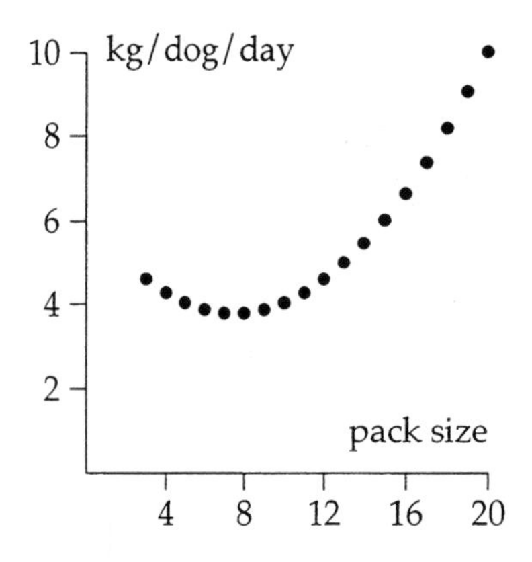

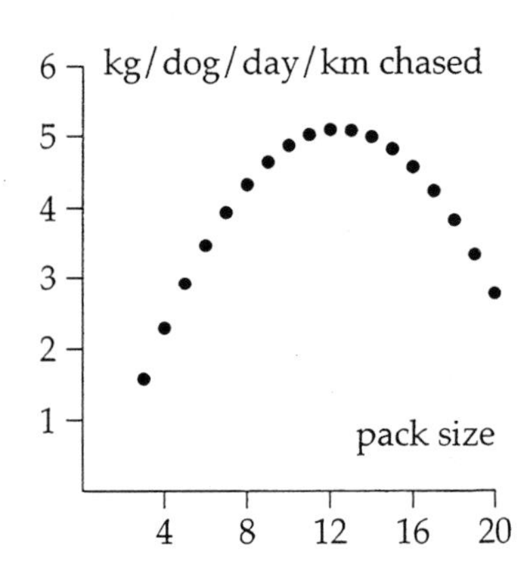

Figure 7. Pack size plotted against kg prey killed per dog per day **(a)** and kg prey killed per dog per day per km of chase **(b)**. See Creel and Creel [1995].

8.3 Conclusion

Optimal foraging theory has been used in several experiments to predict the foraging behavior of animals. We have seen that the model can be and has been adapted in a variety of ways to predict the behavior of animals foraging in a patchy environment or foraging in a group.

But it is also the case that the examples, experiments, and special cases briefly described in this last section begin to make one think that the entire theory is a bit circular. That is, are we simply starting with the premise that animals forage optimally and it is simply a matter of finding the correct "currency" that they are optimizing? Great tits minimize commuting energy, while wild dogs hunting in packs seem to maximize the quantity kilograms of prey killed per dog per day per kilometer chased. Are we simply testing the ingenuity of researchers in designing experiments and interpreting data they collected? In fact, whenever a test of optimal foraging fails, it is always possible that it was because the wrong "currency" was used. For example, in the redshank experiment, should the currency have involved nutrients?

Some would also argue that evolution cannot produce "optimal" foragers, because environmental conditions are constantly changing. But there is always a lag in the evolutionary response to a changing environment.

Others criticize optimal foraging theory as overly simplistic. They argue that making net caloric intake rate the evolutionary currency of the model ignores factors like the need to avoid predators and the constraints of the way genetics and behavior interrelate. See Beardsley [1988] and Krebs [1978] for a further discussion of these points.

Nonetheless, there are an impressive number of experiments that provide supporting evidence for optimal foraging theory, and it remains an active area of continued research. Additional information at an introductory level may be found in Krebs and Davies [1978] and in Lendrem [1986].

9. Solutions to Selected Exercises

1. **a)** $1/k$ per sec. **b)** $2/k$ per sec. **c)** $3/k$ per sec.

2. **a)** Delis: $1/12$ per min; knish carts: $1/6$ per min; hotdog carts: $1/2$ per min. **b)** $3/4$ per min. **c)** $2/3$ per min.

3. **b)** Profitabilities were 0.68 g/s, 1.73 g/s, and 8.41 g/s, respectively. **c)** Handling times were 17.3 s and 10.7 s, respectively.

4. $A = \lambda_1 E_1 + \lambda_2 E_2$, $B = 1 + \lambda_1 h_1 + \lambda_2 h_2$, $a = E_3$, and $b = h_3$.

5. **b)** $\lambda_L \approx 0.093$ large insects per second, or 5.6 per minute.

6. Only Diets 2 and 3 satisfy the "all or nothing" rule of optimal forging. So Diets 1, 4, and 5 are not possible.

7. **a)** Profitabilities (kcal/min) for mussels: 25; shore worms: 120; crabs: 60. **b)** Optimal diet: worms and crabs: ~ 43.6 kcal/min. **c)** Optimal diet: worms and large mussels: ~ 95.5 kcal/min.

9. **a)** Worms alone: 24 kcal/min; crabs alone: 36 kcal/min.

10. **a)** $\lambda_1 > \dfrac{E_2}{h_2 E_1 - h_1 E_2}$.

11. **a)** Summer diet: caterpillars and berries; 130 kcal/min. **b)** Autumn diet: caterpillars, berries, and seeds; ~ 39.7 kcal/min.

12. **a)** From **Exercise 10**, $\lambda_1 > 2.\overline{2}$.

b) The intake rate for worms and moths must be less than the profitability of grubs, or

$$\frac{0.1(162) + \lambda_2(24)}{1 + 0.1(3.6) + \lambda_2(0.6)} < 25 \iff \lambda_2 < 1.9\overline{7}.$$

13. **a)** $E = 0.024L^{3.47} - 0.936L^{1.58} - 362$ joules. **c)** 38 mm. **e)** Clams 29 mm and larger; 40.19 j/s.

15. The number of medium *Daphnia* would be $20/0.71 \sim 28$, and the number of small would be $20/0.27 \sim 74$.

16. **a)** As described in **Section 4**, $\lambda > \dfrac{E_2}{h(E_1 - E_2)} = 1.$

b) This time $\lambda > \dfrac{E_3}{h(E_1 + 0.75E_2 - 1.75E_3)} \approx 0.59.$

c) Now $\lambda > \dfrac{E_4}{h(E_1 + 0.75E_2 + 0.50E_3 - 2.25E_4)} \approx 0.25.$

d) $\lambda < \dfrac{E_4}{h(E_1 + E_2 + E_3 - 3E_4)} \approx 0.22.$

17. **a)** Profitability is $P(x) = 0.000031x^{2.624}/(0.927x - 2.71)$ gm/s, where x is the length in mm.

b) $P'(x)$ is positive on the interval, so the maximum occurs at the endpoint $x = 40$ mm.

18. The optimal diet without recognition time is crabs and shore worms: ~ 30.8 kcal/min. The optimal diet with recognition time is crabs, shore worms, and large snails: ~ 28.3 kcal/min. The diet breadth widens and the net intake rate drops when recognition time is accounted for.

19. With butterflies included, but ignoring recognition time, the optimal diet would be viceroys, worms, and moths, with an intake rate of about 32.5 kcal/min. This is higher than originally, so the habitat has been "enriched" by the inclusion of butterflies. The intake rate is also higher than if recognition time is included and the diets are different.

21. **a)** The critical value is still $T = a$, so the patch residence time does not change. **b)** All the patches have remained relatively the same.

References

Beardsley, T. 1988. A philosophical dispute raises hackles among biologists. *Scientific American* 258 (1): 21–22.

Bryant, D.M., and A.K. Turner. 1982. Central place foraging by swallows (*Hirundinidae*): The question of load size. *Animal Behavior* 30: 845–856.

Charnov, E.L. 1976. Optimal foraging: attack strategy of a mantid. *American Naturalist* 110: 141–151.

Cowie, R.J. 1977. Optimal foraging in great tits (*Parus major*). *Nature* 268: 137–139.

Creel, S., and N.M. Creel. 1995. Communal hunting and pack size in African wild dogs, *Lyacon pictus*. *Animal Behavior* 50: 1325–1339.

Davies, N.B. 1977. Prey selection and social behavior in wagtails (*Aves: Motacilladae*). *Journal of Animal Ecology* 46: 37–57.

De Benedictis, P.A., et al. 1978. Optimal meal size in hummingbirds. *American Naturalist* 112: 301–316.

Elner, R.W., and R.N. Hughes. 1978. Energy maximisation in the diet of the Shore Crab, *Carcinus Maenas (L.)*. *Journal of Animal Ecology* 47: 103–116.

Evans, P.R. 1976. Energy balance and optimal foraging strategies in shorebirds: some implications for their distributions and movements in the nonbreeding season. *Ardea* 64: 117–139.

Giraldeau, L., D.L. Kramer, I. Deslandes, and H. Lair. 1994. The effect of competitors and distance on central place foraging eastern chipmunks, *Tamias striatus*. *Animal Behavior* 47: 621–632.

Goss-Custard, J.D. 1977. Optimal foraging and the size selection of worms by redshank *Tringa totanus*. *Animal Behavior* 25: 10–29.

Jenkins, S.H. 1980. A size–distance relation in food selection by beavers. *Ecology* 61: 740–746.

Kaufman, J.D, G.M. Burghardt, and J.A. Phillips. 1996. Sensory cues and foraging decisions in a large carnivorous lizard, *Varanus albigularis*. *Animal Behavior* 52: 727–736.

Kolmes, S.A. 1990. Recent progress in the study of adaptive behavioural flexibility in honeybees. *Bee World* 71: 122–129.

Krebs, John R. 1978. Optimal foraging: Decision rules for predators. In Krebs and Davies [1978], 23–63. Sunderland, MA: Sinauer Associates.

________, and N.B. Davies. 1978. *Behavioural Ecology, An Evolutionary Approach*. Sunderland, MA: Sinauer Associates.

Lendrem, Dennis. 1986. *Modelling in Behavioral Ecology: An Introductory Text*. 35–57. Portland, OR: Timber Press.

Pitelka, F.A., P.Q. Tomrich, and G.W. Treichel. 1955. Ecological relations of jaegers and owls as lemming predators near Barrow, Alaska. *Ecological Monographs* 25: 85–117.

Richardson, H., and N.A.M. Verbeek. 1986. Diet selection and optimization by Northwestern crows feeding on Japanese littleneck clams. *Ecology* 67: 1219–1226.

Stephens, David W., and John R. Krebs. 1986. *Foraging Theory*. Princeton, NJ: Princeton University Press.

Sutherland, W.J. 1982. Do oystercatchers select the most profitable cockles? *Animal Behavior* 30: 857–861.

Thompson, D.B.A., and C.J. Barnard. 1984. Prey selection by plovers: optimal foraging in mixed species groups. *Animal Behavior* 32: 544–563.

Turner, A.K. 1982. Optimal foraging by the swallow (*Hirundo rustica,* L): Prey size selection. *Animal Behavior* 30: 862–872.

Waite, T.A., and R.C. Ydenberg. 1996. Foraging currencies and the load-size decision of scatter-hoarding grey jays. *Animal Behavior* 51: 903–916.

Werner, E.E., and D.J. Hall. 1974. Optimal foraging and the selection of prey by the bluegill sunfish *Lepomis macrochirus*. *Ecology* 55: 1042–1052.

About the Authors

Steven Kolmes received his B.S. in zoology from Ohio University and his M.S. and Ph.D. degrees in zoology from the University of Wisconsin at Madison. He currently holds the Rev. John Molter, C.S.C., Chair in Science at the University of Portland. He is interested in behavioral ecology at the pest–pesticide interface and efficiency theory in social insects.

Kevin Mitchell received his B.A. in mathematics and philosophy from Bowdoin College and his Ph.D. in mathematics from Brown University. His main areas of interest are hyperbolic tilings, algebraic geometry, and applications of mathematics to environmental science.

James Ryan received his B.A. in zoology from SUNY College at Oswego, NY, his M.S. in biological sciences from the University of Michigan, and his Ph.D. from the University of Massachusetts, Amherst. His main areas of interest are functional morphology and ecology of vertebrates and the evolution of vertebrate neural control systems.

This Module was developed by the authors at Hobart and William Smith Colleges for a team-taught interdisciplinary course entitled "Mathematical Models and Biological Systems."

UMAP

Modules in Undergraduate Mathematics and Its Applications

Published in cooperation with

The Society for Industrial and Applied Mathematics,

The Mathematical Association of America,

The National Council of Teachers of Mathematics,

The American Mathematical Association of Two-Year Colleges,

The Institute for Operations Research and the Management Sciences, and

The American Statistical Association.

Module 765

How Does the NFL Rate Passers?

Roger W. Johnson

Applications of Linear Algebra to Fitting Least Squares and Elementary Perturbation Theory

COMAP, Inc., Suite 210, 57 Bedford Street, Lexington, MA 02173 (781) 862–7878

INTERMODULAR DESCRIPTION SHEET: UMAP Unit 765

TITLE: How Does the NFL Rate Passers?

AUTHOR: Roger W. Johnson
Dept. of Mathematics and Computer Science
South Dakota School of Mines and Technology
501 East St. Joseph Street
Rapid City, SD 57701–3995
rwjohnso@silver.sdsmt.edu
http://silver.sdsmt.edu/~rwjohnso

MATHEMATICAL FIELD: Linear algebra

APPLICATION FIELD: Fitting by least squares, elemenuary perturbation theory

TARGET AUDIENCE: Students who have had introductory courses in differential calculus and elementary matrix algebra

ABSTRACT: The U.S. National Football League rates passers by their performance in four areas: percentage of pass completions, percentage of touchdown passes, percentage of interceptions, and average gain per pass attempt. Such ratings are published in a number of newspapers during the football season, including *USA Today*. Exactly how the NFL computes this rating, however, is not well publicized. In this Module, we see that the NFL rating is a weighted average of the four statistics indicated, and we recover these weights.

PREREQUISITES: Ability to differentiate polynomials and perform basic matrix manipulations. A few exercises require the use of software that performs basic matrix calculations.

COMAP, Inc., Suite 210, 57 Bedford Street, Lexington, MA 02173
(800) 77-COMAP = (800) 772-6627, or (781) 862-7878; http://www.comap.com

How Does the NFL Rate Passers?

Roger W. Johnson
Dept. of Mathematics and Computer Science
South Dakota School of Mines and Technology
501 East St. Joseph Street
Rapid City, SD 57701–3995
rwjohnso@silver.sdsmt.edu
http://silver.sdsmt.edu/~rwjohnso

Table of Contents

Modules and Monographs in Undergraduate Mathematics and its Applications (UMAP) Project

The goal of UMAP is to develop, through a community of users and developers, a system of instructional modules in undergraduate mathematics and its applications, to be used to supplement existing courses and from which complete courses may eventually be built.

The Project was guided by a National Advisory Board of mathematicians, scientists, and educators. UMAP was funded by a grant from the National Science Foundation and now is supported by the Consortium for Mathematics and Its Applications (COMAP), Inc., a nonprofit corporation engaged in research and development in mathematics education.

Paul J. Campbell	Editor
Solomon Garfunkel	Executive Director, COMAP

1. Best Career NFL Ratings

The National Football League (NFL) computes a "passer rating" or "passer efficiency" number for each passer. According to Famighetti [1996, 878], this rating is a combined measure over four areas: percentage of pass completions, percentage of touchdown (TD) passes, percentage of interceptions, and average gain per pass attempt. A number of newspapers, including *USA Today*, list the current season's rating for passers. Career ratings for top-rated passers are given in several annual almanacs (e.g., Marshall [1997, 157]). Data for the passers with the highest career rating through the 1996 season, and who have attempted at least 1,500 passes, are given in **Table 1**.

Table 1.

Best career NFL passers by NFL rating.

Player	Attempts	Completions	Yds	TDs	Interceptions	Rating
Steve Young	3192	2059	25479	174	85	96.2
Joe Montana	5391	3409	40551	273	139	92.3
Brett Favre	2693	1667	18724	147	79	88.6
Dan Marino	6904	4134	51636	369	209	88.3
Otto Graham	2626	1464	23584	174	135	86.6
Jim Kelly	4779	2874	35467	237	175	84.4
Roger Staubach	2958	1685	22700	153	109	83.4
Troy Aikman	3178	2000	22733	110	98	83.0
Neil Lomax	3153	1817	22771	136	90	82.7
Sonny Jurgensen	4262	2433	32224	255	189	82.6
Len Dawson	3741	2136	28711	239	183	82.6
Jeff Hostetler	2194	1278	15531	89	61	82.1
Ken Anderson	4475	2654	32838	197	160	81.9
Bernie Kosar	3365	1994	23301	124	87	81.8
Danny White	2950	1761	21959	155	132	81.7
Dave Krieg	5288	3092	37946	261	199	81.5
Warren Moon	6000	3514	43787	254	208	81.0
Neil O'Donnell	2059	1179	14014	72	46	80.5
Scott Mitchell	1507	853	10516	71	49	80.5
Bart Starr	3149	1808	24718	152	138	80.5
Ken O'Brien	3602	2110	25094	128	98	80.4
Fran Tarkenton	6467	3686	47003	342	266	80.4

The particular way in which the NFL computes rating, however, is generally unknown, despite the fact that rating is often discussed by broadcasters, sports writers, and fans. Rating is also used in contract negotiations. According to NFL player's negotiator Robert Fayne [Ellis 1993], "the one thing used more than any other [in quarterback contract negotiations] is that rating." Trade clauses have also involved rating values. For example, Cincinnati was to receive a 1994 second-round draft choice from the New York Jets if Boomer Esiason, traded from Cincinnati to the New York Jets, had a rating that was 89 or higher at the end of the 1993 season.

So how is the NFL rating computed? In this Module, which is an expanded version of Johnson [1993; 1994], our goal is to uncover the formula used.

2. A Simple Approach to Uncovering the Rating Formula

While it is by no means the case that rating is necessarily linear in the four statistics mentioned previously, a natural starting point is to assume such a linear rating formula and see if this linear model can be shown to fit the data in **Table 1**. In particular, suppose that

$$\text{Rating} = \beta_0 + \beta_1\,(\text{Completion }\%) + \beta_2\,(\text{Yards/Attempts}) + \beta_3(\text{Touchdown }\%) + \beta_4\,(\text{Interception }\%)$$

for some unknown values of β_i. If we had the *exact* values of rating and the four passing statistics for five players, then we could solve a system of five equations in five unknowns to determine the coefficients. In particular, using the data in **Table 1** for Montana, Marino, Kosar, Moon, and Tarkenton (some of the author's favorite quarterbacks!), we could set

$$\begin{bmatrix} 1 & 100*\dfrac{3409}{5391} & \dfrac{40551}{5391} & 100*\dfrac{273}{5391} & 100*\dfrac{139}{5391} \\ 1 & 100*\dfrac{4134}{6904} & \dfrac{51636}{6904} & 100*\dfrac{369}{6904} & 100*\dfrac{209}{6904} \\ 1 & 100*\dfrac{1994}{3365} & \dfrac{23301}{3365} & 100*\dfrac{124}{3365} & 100*\dfrac{87}{3365} \\ 1 & 100*\dfrac{3514}{6000} & \dfrac{43787}{6000} & 100*\dfrac{254}{6000} & 100*\dfrac{208}{6000} \\ 1 & 100*\dfrac{3686}{6467} & \dfrac{47003}{6467} & 100*\dfrac{342}{6467} & 100*\dfrac{266}{6467} \end{bmatrix} \begin{bmatrix} \beta_0 \\ \beta_1 \\ \beta_2 \\ \beta_3 \\ \beta_4 \end{bmatrix} = \begin{bmatrix} 92.3 \\ 88.3 \\ 81.8 \\ 81.0 \\ 80.4 \end{bmatrix} \qquad \textbf{(1)}$$

Solving **(1)** gives

$$\beta = [\beta_0, \beta_1, \beta_2, \beta_3, \beta_4]^T = [-0.400,\ 0.884,\ 4.032,\ 3.296,\ -3.974]^T. \qquad \textbf{(2)}$$

The problem with this approach is that ratings are given only to the nearest tenth, so that if the linear model is correct, our solution to **(1)** gives only approximate values of the coefficients. (Some publications, such as Carter and Sloan [1996, 330], give additional decimal digits, to distinguish among players who have, to the nearest tenth, the same rating.) Calling the 5-by-5 matrix in **(1)** A, a more precise way of writing **(1)** would be

$$A(\beta + \Delta\beta) = r + \Delta r, \qquad \textbf{(3)}$$

where $r + \Delta r = [92.3,\ 88.3,\ 81.8,\ 81.0,\ 80.4]$ is the vector of true ratings r (unknown) plus an error vector Δr, and $\beta + \Delta\beta$ is the true vector β of coefficients in our linear model plus an error vector $\Delta\beta$ brought about by the presence of Δr. Consequently, **(2)** is more properly written as

$$\beta + \Delta\beta = [-0.400,\ 0.884,\ 4.032,\ 3.296,\ -3.974]^T. \tag{4}$$

Note that because ratings are rounded to the nearest tenth, the components of the vector Δr are at most 0.05 in absolute value. Using the standard mathematical notation

$$||x||_\infty = \max_i |x_i|,$$

for a vector x, we can write this condition as $||\Delta r||_\infty \leq 0.05$.

At this point we have an estimate of β, namely, the value of $\beta + \Delta\beta$ given in **(4)**. We would like to see how big the error in our estimate, namely $\Delta\beta$, can be. To this end, define $||A||_\infty$ for a matrix A as

$$||A||_\infty \equiv \max_{x \neq 0} \frac{||Ax||_\infty}{||x||_\infty}.$$

Note that we are defining $||\cdot||$ for a matrix argument in terms of $||\cdot||$ of vector arguments, which has already been defined. Given an arbitrary nonzero vector y, then

$$||A||_\infty \geq \frac{||Ay||_\infty}{||y||_\infty}, \qquad \text{or} \qquad ||Ay||_\infty \leq ||A||_\infty ||y||_\infty.$$

From **(3)** and the fact $A\beta = r$, we have

$$A\Delta\beta = \Delta r, \qquad \text{or} \qquad \Delta\beta = A^{-1}\Delta r,$$

from which we get

$$||\Delta\beta||_\infty = ||A^{-1}\Delta r||_\infty \leq ||A^{-1}||_\infty ||\Delta r||_\infty \leq 0.05||A^{-1}||_\infty. \tag{5}$$

It can be shown (see **Exercise 3**) that for a matrix B with n columns, we have

$$||B||_\infty = \max_i \sum_{j=1}^{n} |b_{ij}|, \tag{6}$$

where b_{ij} is the element in the ith row and jth column of B. Computing A^{-1} and taking the absolute values of the elements, we find the resulting maximum row sum, $||A^{-1}||_\infty$, to be 68.0223, so that

$$||\Delta\beta||_\infty \leq 0.05 \times 68.0223 \approx 3.4.$$

That is, a bound on the maximal absolute error in any of the five estimates in **(2)** is about 3.4. Note that this bound is large compared to the estimated components of β in **(2)**; so, at this point, we are not very confident in the estimate of β given in **(2)**!

Exercises

1. Show that the linear model

$$\text{Rating} = \beta_0 + \beta_1\,(\text{Completion }\%) + \beta_2\,(\text{Yards/Attempts}) \\ + \beta_3(\text{Touchdown }\%) + \beta_4\,(\text{Interception}\%)$$

with β as given in **(2)** correctly produces (up to rounding) the ratings in **Table 1** for Montana, Marino, Kosar, Moon, and Tarkenton. How well does this model work for other quarterbacks in **Table 1**? Compute the above expression for at least three other quarterbacks. Does a linear model appear to be a reasonable one?

2. Consider the system $Ax = b$, where

$$A = \begin{bmatrix} 1 & 1 \\ 1 & 1.01 \end{bmatrix} \qquad \text{and} \qquad b = \begin{bmatrix} 1 \\ 1 \end{bmatrix}.$$

Solve for x. Now suppose that b is observed with error $\Delta b = [0, 0.01]^T$, giving rise to the system $A(x + \Delta x) = b + \Delta b$, that is,

$$\begin{bmatrix} 1 & 1 \\ 1 & 1.01 \end{bmatrix} (x + \Delta x) = \begin{bmatrix} 1 \\ 1.01 \end{bmatrix}.$$

Determine how the introduction of the error changes or "perturbs" the solution x, by computing Δx directly. What is $\|\Delta x\|_\infty$? Also, determine $\|A^{-1}\|_\infty$ and compute the bound $\|A^{-1}\|_\infty \|\Delta b\|_\infty$ (see **(5)**) for $\|\Delta x\|_\infty$.

3. Prove **(6)**. Hint: Note for a constant α and a vector x that $\|\alpha x\|_\infty = |\alpha| \|x\|_\infty$. Consequently,

$$\|B\|_\infty = \max_{x \neq 0} \frac{\|Bx\|_\infty}{\|x\|_\infty} = \max_{x \neq 0} \left\| B \frac{x}{\|x\|_\infty} \right\|_\infty = \max_{\|w\|_\infty = 1} \|Bw\|_\infty.$$

4. Compute A^{-1} using appropriate software and then verify that $\|A^{-1}\|_\infty = 68.0223$ for the selected passers. Also, compute $\|A^{-1}\|_\infty$ and the upper bound $\|A^{-1}\|_\infty \|\Delta r\|_\infty$ for $\|\Delta \beta\|_\infty$ for at least one other selection of five quarterbacks from **Table 1**.

We summarize our results to this point. Using data for Montana, Marino, Kosar, Moon, and Tarkenton in **Table 1**, we estimated the vector of coefficients β of the four passing statistics used to produce the rating. (If you haven't already done so, work **Exercise 1** to convince yourself that the rating apparently is linear in the four statistics.) We weren't very confident in our estimate because the bound on the maximal error in the components was large compared to the estimated components of β. So, when trying to find the solution β to the system

$$A\beta = r,$$

whose right-hand side has been modified to give the "perturbed" system

$$A(\beta + \Delta\beta) = r + \Delta r,$$

we are concerned with not just the size of $\Delta\beta$ but the size of $\Delta\beta$ relative to β. One measure of this relative error is the ratio

$$\frac{\|\Delta\beta\|_\infty}{\|\beta\|_\infty}.$$

To get a bound on this quantity, from **(5)** we have

$$\|\Delta\beta\|_\infty \leq \|A^{-1}\|_\infty \|\Delta r\|_\infty. \quad \textbf{(7)}$$

Also, from $A\beta = r$, we have $\|r\|_\infty \leq \|A\|_\infty \|\beta\|_\infty$, or

$$\frac{1}{\|\beta\|_\infty} \leq \frac{\|A\|_\infty}{\|r\|_\infty}. \quad \textbf{(8)}$$

Consequently, from **(7)** and **(8)**, we have

$$\frac{\|\Delta\beta\|_\infty}{\|\beta\|_\infty} \leq \kappa(A)\,\frac{\|\Delta r\|_\infty}{\|r\|_\infty}, \quad \textbf{(9)}$$

where

$$\kappa(A) \equiv \|A\|_\infty \|A^{-1}\|_\infty$$

is the *condition number* of the matrix A. If the condition number of A is "small," then a small relative change in the right-hand side of the system $A\beta = r$ will guarantee a small relative change in the solution. On the other hand, if the condition number of A is "large," then it is possible that a small relative change in the right-hand side will produce a large relative change in the solution. The matrix A is said to be *ill-conditioned* when $\kappa(A)$ is large.

For more details on how the solution to the system is changed with perturbations not only to r but also possibly in A, see Strang [1988, 362–369] and Watkins [1991, 94–109].

Exercises

5. **a)** For A as in **Exercise 2**, show that $\kappa(A) = 404.01$, so that **(9)** becomes

$$\frac{\|\Delta x\|_\infty}{\|x\|_\infty} \leq 404.01\,\frac{\|\Delta b\|_\infty}{\|b\|_\infty}.$$

   b) For A the matrix in **(1)** (see **Exercise 4**), show that $\kappa(A) = 5{,}400.93$, so that **(9)** becomes

$$\frac{\|\Delta\beta\|_\infty}{\|\beta\|_\infty} \leq 5{,}400.93 \times \frac{\|\Delta r\|_\infty}{\|r\|_\infty} \leq 5{,}400.93 \times \frac{0.05}{92.25} \approx 2.93,$$

   or $\|\Delta\beta\|_\infty \leq 2.93\|\beta\|_\infty$, which is not very encouraging.

6. Show that

a) $\kappa(I) = 1$, where I is an identity matrix;

b) $\kappa(A^{-1}) = \kappa(A)$;

c) $\kappa(cA) = \kappa(A)$, where c is a nonzero constant; and

d) $\kappa(A) \geq 1$.

7. a) Show that

$$\kappa(A) \geq \frac{|\lambda_L|}{|\lambda_S|},$$

where λ_L and λ_S are, respectively, the absolute largest eigenvalue and the absolute smallest eigenvalue of A. (Hint: If λ is an eigenvalue of a nonsingular matrix A, then $1/\lambda$ is an eigenvalue of A^{-1}.)

b) Verify that this inequality holds for the matrix A given in **Exercise 2**.

8. If both A and r in the system $A\beta = r$ are observed with error, giving rise to the perturbed system

$$(A + \Delta A)(\beta + \Delta\beta) = r + \Delta r,$$

then (see, for example, Watkins [1991, 106]), assuming

$$\frac{\|\Delta A\|_\infty}{\|A\|_\infty} < \frac{1}{\kappa(A)},$$

it can be shown that

$$\frac{\|\Delta\beta\|_\infty}{\|\beta\|_\infty} \leq \frac{\kappa(A)\left(\dfrac{\|\Delta A\|_\infty}{\|A\|_\infty} + \dfrac{\|\Delta r\|_\infty}{\|r\|_\infty}\right)}{1 - \kappa(A)\dfrac{\|\Delta A\|_\infty}{\|A\|_\infty}}.$$

We apply this inequality to another setting. In an attempt to find the coefficients of the model

Calories = a(g of carbohydrates) + b(g of protein) + c(g of fat),

look at three food labels and record the number of calories, the number of grams of carbohydrates, the number of grams of protein, and the number of grams of fat for each. Using your data,

a) Estimate the coefficients a, b, and c in the model.

b) Make some assumptions about how each of the four values are rounded (e.g., calories are rounded to the nearest 5 calories and the grams of carbohydrates, protein, and fat are each rounded to the nearest gram), and obtain an upper bound for $\|\Delta\beta\|_\infty/\|\beta\|_\infty$ in terms of $\kappa(A)$, assuming that the above inequality holds.

3. A Least-Squares Approach to Uncovering the Formula

Estimates of β obtained by the method presented in **Section 2** are somewhat disconcerting; they don't use all of the data of **Table 1**. We could take all ways of selecting 5 quarterbacks from the list of 22 and estimate β by the value of $\beta + \Delta\beta$ for which the bound **(9)** on $\|\Delta\beta\|_\infty / \|\beta\|_\infty$ is smallest (cf. **Exercise 5b**). Unfortunately, there are $\binom{22}{5} = 26{,}334$ ways of selecting 5 from 22!

An alternative approach, still assuming the linear model, is to try to choose β to make the fitted values of

$$\beta_0 + \beta_1(\text{Completion }\%) + \beta_2(\text{Yards/Attempts}) + \beta_3(\text{Touchdown }\%) + \beta_4(\text{Interception }\%)$$

close to the observed rounded ratings in some collective sense. Let S be the 22-by-5 matrix of player's statistics

$$S = \begin{bmatrix} 1 & 100 * \dfrac{2059}{3192} & \dfrac{25479}{3192} & 100 * \dfrac{174}{3192} & 100 * \dfrac{85}{3192} \\ 1 & 100 * \dfrac{3409}{5391} & \dfrac{40551}{5391} & 100 * \dfrac{273}{5391} & 100 * \dfrac{139}{5391} \\ \vdots & \vdots & \vdots & \vdots & \vdots \\ 1 & 100 * \dfrac{3686}{6467} & \dfrac{47003}{6467} & 100 * \dfrac{342}{6467} & 100 * \dfrac{266}{6467} \end{bmatrix}$$

The first row contains Young's statistics, the second row Montana's, etc. Let P be the vector of length 22 of observed player ratings

$$P = [96.2,\ 92.3,\ \ldots,\ 80.4]^T.$$

Then $(S\beta)_i$ is the rating for player i, and P_i is the associated rounded rating. A reasonable approach to estimating β that is mathematically tractable is to choose β to minimize the sum of squared differences between these two values. That is, choose β to minimize

$$\begin{aligned} &\sum_{i=1}^{22} (\text{Rating Player } i - \text{Observed Rounded Rating Player } i)^2 \\ &\qquad = \sum_{i=1}^{22} [(S\beta)_i - P_i]^2 = (S\beta - P)^T (S\beta - P) \\ &\qquad = \beta^T S^T S\beta - P^T S\beta - \beta^T S^T P + P^T P, \end{aligned}$$

which, because $P^T S\beta = (P^T S\beta)^T = \beta^T S^T P$ (since a scalar and its transpose are equal), becomes

$$= \beta^T S^T S\beta - 2\beta^T S^T P + P^T P.$$

This sum of squared errors depends on β; in what follows, call it $SS(\beta)$. Consequently,

$$SS(\beta) = \beta^T S^T S\beta - 2\beta^T S^T P + P^T P, \tag{10}$$

and we can find the optimal "least squares" components of β by setting

$$\frac{\partial SS(\beta)}{\partial \beta_0} = \frac{\partial SS(\beta)}{\partial \beta_1} = \cdots = \frac{\partial SS(\beta)}{\partial \beta_4} = 0. \tag{11}$$

A more compact way of writing **(11)** is

$$\frac{\partial SS(\beta)}{\partial \beta} = 0,$$

where

$$\frac{\partial SS(\beta)}{\partial \beta} = \begin{bmatrix} \frac{\partial SS(\beta)}{\partial \beta_0} \\ \vdots \\ \frac{\partial SS(\beta)}{\partial \beta_4} \end{bmatrix}$$

and $0 = [0,0,0,0,0]^T$. More generally, if f is a real-valued function of $\beta = [\beta_0, \ldots, \beta_n]^T$, then define

$$\frac{\partial f(\beta)}{\partial \beta} \equiv \begin{bmatrix} \frac{\partial f(\beta)}{\partial \beta_0} \\ \vdots \\ \frac{\partial f(\beta)}{\partial \beta_n} \end{bmatrix}. \tag{12}$$

Exercises

9. If $f(\beta_0, \beta_1, \beta_2) = 5\beta_0 - 2\beta_1 + 7\beta_2$, show that

$$\frac{\partial f}{\partial \beta} = \begin{bmatrix} 5 \\ -2 \\ 7 \end{bmatrix}.$$

More generally, if $f(\beta) = \beta^T c$, show that

$$\frac{\partial f}{\partial \beta} = c.$$

10. If $\beta = [\beta_0, \beta_1, \beta_2]^T$ and

$$Q = \begin{bmatrix} 1 & 2 & 3 \\ 2 & 4 & 5 \\ 3 & 5 & 6 \end{bmatrix},$$

compute $f(\beta) = \beta^T Q \beta$ and $\partial f / \partial \beta$. Also, compute $2Q\beta$ and show that it equals $\partial f / \partial \beta$.

11. For $f(\beta) = \beta^T Q \beta$ with Q a symmetric matrix, verify that

$$\frac{\partial}{\partial \beta} \left(\beta^T Q \beta\right) = 2Q\beta.$$

12. Show that for any $(n+1) \times (n+1)$ matrix M,

$$\beta^T M \beta = \beta^T \left(\frac{M + M^T}{2}\right) \beta.$$

What does this say about the generality of the result in **Exercise 11**? Compute $\partial f / \partial \beta$ for $f(\beta) = \beta^T Q \beta$, where $\beta = [\beta_0, \beta_1, \beta_2]^T$ and

$$Q = \begin{bmatrix} 1 & 2 & 3 \\ 4 & 5 & 6 \\ 7 & 8 & 9 \end{bmatrix}.$$

From one-variable calculus, we know that

$$\frac{d}{dx}(cx) = c \qquad \text{and} \qquad \frac{d}{dx}\left(cx^2\right) = 2cx;$$

for the vector derivative defined in **(12)**, we have, from **Exercises 9** and **11**,

$$\frac{\partial}{\partial \beta}\left(\beta^T c\right) = c \qquad \text{and} \qquad \frac{\partial}{\partial \beta}\left(\beta^T Q \beta\right) = 2Q\beta \qquad \text{for symmetric } Q.$$

Applying these two results to finding the optimal least squares value of β, denoted $\hat{\beta}_{\text{LS}}$, we have from **(10)**

$$0 = \frac{\partial SS(\beta)}{\partial \beta} = \frac{\partial \left(\beta^T S^T S \beta - 2\beta^T S^T P + P^T P\right)}{\partial \beta} = 2S^T S \beta - 2S^T P,$$

(note that $S^T S$ is symmetric), which for invertible $S^T S$ implies that

$$\hat{\beta}_{\text{LS}} = \left(S^T S\right)^{-1} S^T P. \qquad \textbf{(13)}$$

For the S and P associated with **Table 1**, we have

$$\hat{\beta}_{\text{LS}} = [2.035,\ 0.835,\ 4.157,\ 3.334,\ -4.159]^T. \qquad \textbf{(14)}$$

Note that the estimates of β given in **(2)** and **(14)** agree reasonably well except for the estimated values of β_0.

We summarize: To fit $S\beta$ to P for a given matrix S and a given vector P, the optimal β (in terms of least squares) is given by **(13)**.

Exercise

13. Use **(13)** to:

- **a)** Find the best-fitting (least-squares) line to the points $(-2, 1)$, $(0, 3)$, $(3, 5)$. Also, plot the three points and the least-squares line.
- **b)** Likewise, find the best-fitting (least-squares) line through the origin to the points $(-1, -1)$ and $(1, 2)$. (Hint: The S matrix is actually a column vector in this case, a column that is not entirely composed of 1s.) Also, plot the two points along with the least-squares line through the origin.

The fitted values of

$$\beta_0 + \beta_1\,(\text{Completion }\%) + \beta_2\,(\text{Yards/Attempts}) + \beta_3\,(\text{Touchdown }\%) + \beta_4(\text{Interception }\%)$$

with $\beta = \hat{\beta}_{\text{LS}}$ are very close to the values of rounded rating given in **Table 1**, the differences being less than 0.05 for all quarterbacks except Dan Marino, where the error is about 0.0513 (the method of least squares does not guarantee that all these differences be less than 0.05). So, the linear model assumption appears to be a reasonable one. Assuming that the NFL uses "nice" rational values for the components of β, a little bit of trial and error suggests the formula

$$\text{Rating} = \frac{25 + 10(\text{Compl. }\%) + 50(\text{Yds/Attempts}) + 40(\text{TD }\%) - 50(\text{Interc. }\%)}{12}, \qquad \textbf{(15)}$$

(which corresponds to $\beta = [2.08\overline{3},\ 0.8\overline{3},\ 4.1\overline{6},\ 3.\overline{3},\ -4.1\overline{6}]$; cf. **(14)**). This formula does indeed produce the ratings listed in **Table 1** upon rounding to the nearest tenth.

4. Models and Extrapolation

The true test of any model is to see how it performs on data that were not used to build it. Consequently, to confirm our belief of **(15)** as being the method that the NFL uses to rate passers, we should see if it works on data other than those of **Table 1**. We find that it works with nearly all passers. It seems to fail occasionally, for passers who have just a handful of passing attempts. From Hollander [1996, 392–393], who lists passer ratings for all players who attempted at least one pass during the 1995 season, we see that San Francisco's wide receiver Jerry Rice threw one completed pass of 41 yards for a touchdown in 1995. Consequently his 1995 rating, as computed by **(15)**, is

$$\frac{25 + 10(100) + 50(41) + 40(100) - 50(0)}{12} = 589.6.$$

Rice's actual rating for 1995, however, was 158.3. As is the case more generally in modeling problems, it should not be assumed that a model built on a certain kind of data (e.g., passers with quite a few passing attempts) can be applied to other data (e.g., passers with only a limited number of passing attempts). By examining a document available from the NFL [1977], one can conclude that **(15)** agrees with the NFL rating method for those passers who satisfy

$$\begin{array}{rcl} 30\% & \leq \text{Completion } \% & \leq 77.5\% \\ 3.0 & \leq \text{Yards/Attempt} & \leq 12.5 \\ 0\% & \leq \text{Touchdown } \% & \leq 11.875\% \\ 0\% & \leq \text{Interception } \% & \leq 9.5\%. \end{array}$$

Such inequalities hold for nearly all passers. The actual rating method, not stated as such, but which we conclude from the variety of tables listed in NFL [1977], is

$$\begin{aligned} \text{NFL Rating} = {} & \left[\frac{5}{6}(\text{Completion } \% - 30)\right] + \left[\frac{25}{6}(\text{Yards/Attempts} - 3)\right] \\ & + \left[\frac{10}{3}(\text{Touchdown } \%)\right] + \left[\frac{25}{12}(19 - 2(\text{Interception } \%))\right], \end{aligned} \qquad \textbf{(16)}$$

where it is understood that any value in square brackets is truncated to be no smaller than zero and no larger than 475/12. This implies a minimal rating of 0 and a maximal rating of $158.\overline{3}$. The NFL, by the way, has used a variety of techniques to rate passing since the 1932 season. From 1932 to 1937, for instance, passers were ranked by total yards passing. From 1938 to 1940, passers were ranked by the percentage of completions. A variety of other, generally more complicated ranking schemes followed in subsequent years, with the current system being adopted in 1973.

Exercises

14. Verify that **(16)** reduces to **(15)** when the above inequalities hold.

15. A number of folks have criticized the NFL passer-rating method. Proposals for alternative methods have even appeared in the *Wall Street Journal* [Barra and Neyer 1995]. Harvard statistician Carl Morris notes the following defect with the NFL rating: Most passers would improve their rating by completing an extra pass for no gain. Verify that this claim is true for any passer with a rating less than $83.\overline{3}$ (assume that the necessary conditions for **(16)** to reduce to **(15)** hold).

16. National Collegiate Athletic Association (NCAA) passers are also ranked by a rating formula, but the formula is different from that of the NFL. The relevant statistics on 13 players are given in **Table 2** (see Bollig [1991]).

Assuming that the NCAA uses a linear combination of the same four passing statistics as the NFL, estimate, using appropriate software, what

Table 2.
Passing statistics for some NCAA quarterbacks.

Player	Attempts	Completions	Yds	TDs	Intercepts	NCAA Rating
John Elway	1246	774	9349	77	39	139.3
Jim Everett	923	550	7158	40	30	132.5
Doug Flutie	1270	677	10579	67	54	132.2
Bert Jones	418	221	3255	28	16	132.7
Tommy Kramer	1036	507	6197	37	52	100.9
Archie Manning	761	402	4753	31	40	108.2
Jim McMahon	1060	653	9536	84	34	156.9
Joe Montana	515	268	4121	25	25	127.3
Rodney Peete	972	571	7640	52	32	135.8
Vinny Testaverde	674	413	6058	48	25	152.9
Joe Theismann	509	290	4411	31	35	136.1
Andre Ware	1074	660	8202	75	28	143.4
Steve Young	908	592	7733	56	33	149.8

the NCAA rating formula is. (A partial rationale for the NCAA rating method is given in Bollig [1991, 4] and in Summers [1993, 8]; apparently, there is no published rationale for the NFL method.) Also, if the NFL were to adopt the NCAA rating method, how would the order of the 22 quarterbacks in **Table 1** change? The Canadian Football League and World League of American Football, by the way, apparently use the same passer rating method as the NFL.

5. Concluding Comments

In most instances, the NFL rating for a passer is a linear combination of the four passing statistics of completion percentage, yards per attempt, touchdown percentage, and interception percentage. We estimated the coefficients in such a linear model by two different methods. The least-squares approach of **Section 3** is preferred to the simple approach of **Section 2**, since the former makes use of all the data. There are, however, still other ways to proceed. We conclude by briefly discussing two alternative approaches to least squares.

Instead of minimizing the sum of squared errors, an alternative approach would be to minimize the sum of the absolute errors. That is, find β to minimize

$$\sum_{i=1}^{n} |P_i - (S\beta)_i|,$$

where n is the number of players. Another alternative approach would be to find β to minimize

$$\max_i |P_i - (S\beta)_i|.$$

Interestingly enough, each of these two alternative approaches can be cast as *linear programming* problems—problems in which one optimizes a linear function subject to linear inequality constraints, for which efficient computer code has been written.

This second alternative approach, for instance, may be posed as the following linear programming problem:

$$\begin{aligned} &\text{minimize } z \\ &\text{subject to} \\ z + (S\beta)_i &\geq P_i, \qquad i = 1, 2, \dots, n \\ z - (S\beta)_i &\geq -P_i, \qquad i = 1, 2, \dots, n. \end{aligned}$$

To see why this is the case, note that each pair of inequality constraints is equivalent to

$$z \geq |P_i - (S\beta)_i|.$$

So, for any optimal solution $(z, \beta) = (z^*, \beta^*)$, the optimal z^* is slammed down against the largest of the n bounds $|P_i - (S\beta^*)_i|$, giving

$$z^* = \max_i |P_i - (S\beta^*)_i|,$$

and β^* is a desired optimal choice for β. See, for example, Chvátal [1983, 221–227] for further information on both alternative approaches.

Linear programming problems allow for the inclusion of inequality constraints. If we were to adopt the first alternative approach, say, we could impose the linear inequality constraints that the fitted ratings match the observed ratings, upon rounding, to the nearest tenth. Note that such constraints would not be needed with the second alternative approach, unless, as in Carter and Sloan [1996, 330], the ratings were given to different accuracies. At any rate, these last two alternative approaches could be implemented in such a way that the fitted ratings matched the observed ratings up to the necessary number of decimal digits. Such correctness, up to rounding, is not guaranteed by the least-squares approach of **Section 3**.

6. Solutions to the Exercises

1. Using the linear model with β as in **(2)** gives a fairly good job on all 22 quarterbacks; a linear model seems reasonable. For 12 of the 22 quarterbacks, the computed rating, when rounded to the nearest tenth, disagrees with the rating reported in **Table 1**. The worst error, of about 0.256, occurs with Troy Aikman.

2. Solving the two systems of equations we find that $x = [1, 0]^T$ and $x + \Delta x = [0, 1]^T$, so that $\Delta x = [-1, 1]^T$ and $\|\Delta x\|_\infty = 1$. As

$$A^{-1} = \begin{bmatrix} 101 & -100 \\ -100 & 100 \end{bmatrix},$$

we have $\|A^{-1}\|_\infty = 201$ and $\|\Delta x\|_\infty \le \|A^{-1}\|_\infty \|\Delta b\|_\infty \le (201)(0.01) = 2.01$.

3. Continuing from the hint, note that $(Bw)_i = \sum_{j=1}^n b_{ij} w_j$ and that the maximum over $\|w\|_\infty = 1$ occurs when the components of w are ± 1 according to the signs of the b_{ij}.

6. **a)** $\kappa(I) = \|I\|_\infty \|I^{-1}\|_\infty = \|I\|_\infty \|I\|_\infty = 1 \cdot 1 = 1$.

b) $\kappa(A^{-1}) = \|A^{-1}\|_\infty \|(A^{-1})^{-1}\|_\infty = \|A^{-1}\|_\infty \|A\|_\infty = \kappa(A)$.

c) $\kappa(cA) = \|cA\|_\infty \|(cA)^{-1}\|_\infty = \|cA\|_\infty \|(1/c)A\|_\infty = |c| \|A\|_\infty |1/c| \|A^{-1}\|_\infty$ $= \|A\|_\infty \|A^{-1}\|_\infty = \kappa(A)$.

d) $1 = \|I\|_\infty = \|AA^{-1}\|_\infty \le \|A\|_\infty \|A^{-1}\|_\infty = \kappa(A)$.

7. **a)** As $Ax_L = \lambda_L x_L$, it follows that

$$\|A\|_\infty = \max_{x \neq 0} \frac{\|Ax\|_\infty}{\|x\|_\infty} \ge \frac{\|Ax_L\|_\infty}{\|x_L\|_\infty} = \frac{\|\lambda_L x_L\|_\infty}{\|x_L\|_\infty} = |\lambda_L| \frac{\|x_L\|_\infty}{\|x_L\|_\infty} = |\lambda_L|.$$

From $Ax_S = \lambda_S x_S$, we have $x_S = A^{-1}\lambda_S x_s$, or $A^{-1}x_S = (1/\lambda_S)x_S$, and the same argument as above gives

$$\|A^{-1}\|_\infty \ge \left| \frac{1}{\lambda_S} \right|.$$

Consequently,

$$\kappa(A) = \|A\|_\infty \|A^{-1}\|_\infty \ge \frac{|\lambda_L|}{|\lambda_S|}.$$

b) $404.01 = \kappa(AS) \ge \dfrac{|\lambda_L|}{|\lambda_S|} \approx \dfrac{2.0050125}{0.0049875} \approx 402.0075188.$

8. Answers will vary with the data collected. With the data of **Table 3**, we have the system

$$\begin{bmatrix} 6 & 5 & 16 \\ 5 & 10 & 15 \\ 11 & 6 & 5 \end{bmatrix} (\beta + \Delta\beta) = (A + \Delta A)(\beta + \Delta\beta) = r + \Delta r = \begin{bmatrix} 180 \\ 200 \\ 120 \end{bmatrix}.$$

Table 3.
Sample data for **Exercise 8**.

Food Item	Calories	Carbohydrates (g)	Protein (g)	Fat (g)
Mixed nuts	180	6	5	16
Peanut butter	200	5	10	15
2% milk	120	11	6	5

a) Solving this system, we find $\beta + \Delta\beta \approx [3.90,\ 6.34,\ 7.80]^T$. According to Brody [1987, 102], for example, $\beta = [4,\ 4, 9]^T$ (cf. Johnson [1995] for estimating β using least-squares as in **Section 3**).

b) From the data collected, it is natural to suppose that calories are rounded to the nearest 10 calories and that grams of carbohydrates, protein, and fat are each rounded to the nearest gram. Consequently,

$$\|\Delta r\|_\infty \leq 5.0, \qquad 195.0 \leq \|r\|_\infty \leq 205.0;$$

and, using **(6)** and noting that the row sums of $(A + \Delta A)$ (a matrix with positive elements) are 27, 30, and 22, we have

$$\|\Delta A\|_\infty \leq 0.5 + 0.5 + 0.5 = 1.5, \qquad 28.5 \leq \|A\|_\infty \leq 30 + 1.5 = 31.5,$$

so

$$\frac{\|\Delta\beta\|_\infty}{\|\beta\|_\infty} \leq \frac{\kappa(A)\left(\dfrac{1.5}{28.5} + \dfrac{5.0}{195.0}\right)}{1 - \kappa(A)\left(\dfrac{1.5}{28.5}\right)}.$$

Note that $\kappa(A) = \|A\|_\infty \|A^{-1}\|_\infty \leq 31.5\|A^{-1}\|_\infty$, but a bound on $\|A^{-1}\|_\infty$ is not easily obtained. One crude approach would be to estimate it by

$$\|A^{-1}\|_\infty \approx \|(A + \Delta A)^{-1}\|_\infty = \left\| \begin{bmatrix} 6 & 5 & 16 \\ 5 & 10 & 15 \\ 11 & 6 & 5 \end{bmatrix}^{-1} \right\|_\infty,$$

which is approximately 0.361.

10.

$$\frac{\partial f}{\partial \beta} = 2Q\beta = 2 \begin{bmatrix} \beta_0 + 2\beta_1 + 3\beta_2 \\ 2\beta_0 + 4\beta_1 + 5\beta_2 \\ 3\beta_0 + 5\beta_1 + 6\beta_2 \end{bmatrix}.$$

11. The "quadratic form" $\beta^T Q\beta$ can be written as

$$\sum_{i=0}^{n}\sum_{j=0}^{n} \beta_i Q_{ij} \beta_j = \sum_{k=0}^{n} Q_{kk}\beta_k^2 + \sum_{i \neq j} \beta_i Q_{ij} \beta_j,$$

so

$$\frac{\partial \beta^T Q \beta}{\partial \beta_t} = 2\beta_t Q_{tt} + \sum_{j \neq t} Q_{tj}\beta_j + \sum_{i \neq t} \beta_i Q_{it} = 2\beta_t Q_{tt} + \sum_{j \neq t} Q_{tj}\beta_j + \sum_{i \neq t} Q_{ti}\beta_i,$$

the last equality from the symmetry of Q. Consequently,

$$\frac{\partial \beta^T Q \beta}{\partial \beta_t} = 2\sum_{i=0}^{n} Q_{ti}\beta_i = 2(Q\beta)_t,$$

and the desired result follows.

13. a) $\sum_i [y_i - (\beta_0 + \beta_1 x_i)]^2 = (S\beta - P)^T(S\beta - P)$, where $P = [1, 3, 5]^T$,

$$S = \begin{bmatrix} 1 & -2 \\ 1 & 0 \\ 1 & 3 \end{bmatrix},$$

and $\beta = [\beta_0, \beta_1]^T$. It follows from **(13)** that $\hat{\beta}_{\text{LS}} = \frac{1}{38}[104, 30]^T$, so that the least-squares line is $y = \frac{1}{38}(104 + 30x)$.

b) $\sum_i [y_i - \beta_0 x_i)]^2 = (S\beta - P)^T(S\beta - P)$, where $P = [-1, 2]^T$, $S = [-1, 1]^T$, and $\beta = [\beta_0]$. It follows from **(13)** that $\hat{\beta}_{\text{LS}} = [1.5]$, so that the least-squares line through the origin is $y = 1.5x$.

14. Note that when the inequalities hold, the [ ] in **(16)** may be removed. The resulting expression simplifies to **(15)**.

15. Suppose that a passer, having thrown A passes, has a quarterback rating of R_C. Let the next pass be completed for no gain, with a resulting rating of R_N. Then, from **(15)**,

$$12R_N(A+1) = 12R_C A + 1000,$$

or

$$R_N = R_C + \frac{1}{A+1}\left(\frac{1000}{12} - R_C\right),$$

so that $R_N > R_C$ when $R_C < 1000/12 = 83.\overline{3}$.

16. According to Hagwell [1993, 8–9], the NCAA passer rating is given by

Completion % + 8.4(Yds/Attempts) + 3.3(TD %) – 2.0(Interception %).

Using computer software, one finds the least-squares fit of rating by

a + b(Completion %) + c(Yds/Attempts) + d(TD %) + e(Interception %)

to be

1.54 + 0.949(Completion %) + 8.73(Yds/Attempts) + 3.16(TD %) – 2.06(Interception %).

If instead one fits rating by

a(Completion %) + b(Yds/Attempts) + c(TD %) + d(Interception %)

(without an additive constant), one obtains

0.972(Completion %) + 8.74(Yds/Attempts) + 3.14(TD %) –2.00(Interception %).

Joe Montana is considered an "outlier" with either model, as the difference between his actual rating and the fitted rating by either model is relatively large. However, the coefficients in either model don't greatly change when we remove Joe's data and fit with the remaining cases. One might say Joe's data does not "outweigh" the other data. In some applications of least squares, the values of $\hat{\beta}_{\text{LS}}$ do substantially change when an observation is omitted. Such observations are said to be *influential*. For further discussion on outliers and influential observations, see, for example, Draper and Smith [1981].

If we apply the NCAA formula, then up to rounding we get the NCAA ratings listed in **Table 2** for all players except for Joe Montana (the NCAA formula gives him a rating of 125.6 instead of 127.3). Presumably, some of the data listed for Joe Montana in Bollig [1991, 160] are in error. It makes sense that observations containing error(s) are often outliers. If an observation is an outlier, however, it doesn't necessarily follow that it is in error.

References

Barra, Allen, and Rob Neyer. 1995. When rating quarterbacks, yards per throw matters. *Wall Street Journal* (24 November 1995): B5.

Bollig, Laura, ed. 1991. *NCAA Football's Finest.* Overland Park, KS: National Collegiate Athletic Association.

Brody, Jane. 1987. *Jane Brody's Nutrition Book.* New York, NY: Bantam Books.

Carter, Craig, and Dave Sloan. 1996. *The Sporting News Pro Football Guide: 1996 Edition.* St. Louis, MO: Sporting News Publishing Company.

Chvátal, Vašek. 1983. *Linear Programming.* New York: W.H. Freeman.

Draper, Norman, and Harry Smith. 1981. *Applied Regression Analysis.* 2nd ed. New York: John Wiley.

Ellis, Elaine. 1993. Mathematics prof. finds NFL secret. *The Carleton Voice* 58 (4):10–11. Northfield, MN: Carleton College.

Famighetti, Robert, ed. 1996. *The World Almanac and Book of Facts 1997.* Mahwah, NJ: K–III Reference Corporation.

Hagwell, Steven, ed. 1993. *1993 Football Statistician's Manual.* Overland Park, KS: National Collegiate Athletic Association.

Hollander, Zander, ed. 1996. *The Complete Handbook of Football: 1996 Edition.* New York: Penguin Books.

Johnson, Roger. 1993. How does the NFL rate the passing ability of quarterbacks? *College Mathematics Journal* 24 (5): 451–453.

_______. 1994. Rating quarterbacks: An amplification. *College Mathematics Journal* 25 (4): 340.

_______. 1995. A multiple regression project. *Teaching Statistics* 17 (2): 64–66.

Marshall, Joe, ed. 1997. *Sports Illustrated 1997 Sports Almanac.* Boston, MA: Little, Brown and Company.

National Football League. 1977. *National Football League Passer Rating System.* New York: NFL.

Strang, Gilbert. 1988. *Linear Algebra and Its Applications.* 3rd ed. San Diego, CA: Harcourt, Brace, Jovanovich.

Summers, J., ed. 1993. *Official 1993 NCAA Football.* Overland Park, KS: National Collegiate Athletic Association.

Watkins, David. 1991. *Fundamentals of Matrix Computations.* New York: John Wiley.

About the Author

Roger Johnson has been at the South Dakota School of Mines & Technology since the fall of 1996 as Associate Professor of Mathematics. For the most part, he teaches courses in probability and statistics, and many of his recent publications concern hands-on learning activities in such. His technical publications include extensions of the James-Stein shrinkage estimator to non-normal settings. When not chasing his children Sarah, 3, and Benjamin, 5, with his wife Laurie he may be found playing softball, running, or foolishly rooting for the Minnesota Vikings.

UMAP

Modules in Undergraduate Mathematics and Its Applications

Published in cooperation with

The Society for Industrial and Applied Mathematics,

The Mathematical Association of America,

The National Council of Teachers of Mathematics,

The American Mathematical Association of Two-Year Colleges,

The Institute for Operations Research and the Management Sciences, and

The American Statistical Association.

Module 766

Using Original Sources to Teach the Logistic Equation

Bonnie Shulman

Applications of Mathematical Modeling to Differential Equations, Biology

COMAP, Inc., Suite 210, 57 Bedford Street, Lexington, MA 02173 (781) 862–7878

INTERMODULAR DESCRIPTION SHEET:	UMAP Unit 766
TITLE:	Using Original Sources to Teach the Logistic Equation
AUTHOR:	Bonnie Shulman Dept. of Mathematics Bates College Lewiston, ME 04240 bshulman@abacus.bates.edu
MATHEMATICAL FIELD:	Differential equations, calculus
APPLICATION FIELD:	Mathematical modeling, biology
TARGET AUDIENCE:	Students in a course in differential equations or second-semester calculus.
ABSTRACT:	This Module uses original data, diagrams, and texts from three original sources to develop the logistic model of growth in natural systems with limited resources. The logistic differential equation and the familiar S-shaped logistic curve have applications in solving problems in ecology, biology, chemistry, and economics. The Module illustrates with concrete examples how mathematics develops, and it provides insights into the assumptions that drive the modeling process.
PREREQUISITES:	The reader is assumed to be familiar with geometric and arithmetic progressions. From calculus: differentiation and integration of elementary functions. A basic introduction to differential equations is desirable, but the Module itself might serve as just such an introduction.

COMAP, Inc., Suite 210, 57 Bedford Street, Lexington, MA 02173
(800) 77-COMAP = (800) 772-6627, or (781) 862-7878; http://www.comap.com

Using Original Sources to Teach the Logistic Equation

Bonnie Shulman
Dept. of Mathematics
Bates College
Lewiston, ME 04240
bshulman@abacus.bates.edu

Table of Contents

Modules and Monographs in Undergraduate Mathematics and its Applications (UMAP) Project

The goal of UMAP is to develop, through a community of users and developers, a system of instructional modules in undergraduate mathematics and its applications, to be used to supplement existing courses and from which complete courses may eventually be built.

The Project was guided by a National Advisory Board of mathematicians, scientists, and educators. UMAP was funded by a grant from the National Science Foundation and now is supported by the Consortium for Mathematics and Its Applications (COMAP), Inc., a nonprofit corporation engaged in research and development in mathematics education.

1. Introduction

There is a common perception of mathematics as a finished product invented by dead geniuses. In an effort to dispel this notion and convey the excitement of mathematics as a living, breathing, and growing body of knowledge, created by human beings very much like ourselves, I have turned to original sources. This Module has grown over time as I integrated material from the three original papers into courses in differential equations, modeling, and even introductory calculus:

- The first paper is an oft-cited classic by Pearl and Reed [1920], who are usually credited with being the first to use the logistic equation to describe the growth of the population of the United States.
- The next paper is the text of a presidential address to the Royal Statistical Society in England, by G. Udny Yule [1925], which contains an excellent critical history of the logistic model and summarizes the work of Pearl and Reed, as well as that of Verhulst.
- From Yule, I learned that Pierre-François Verhulst, a Belgian sociologist and mathematician, was actually the first to propose and publish a formula for the law of growth for a population confined to a specified area [Verhulst 1845].

Yule states, "[p]robably owing to the fact that Verhulst was greatly in advance of his time, and that the then existing data were quite inadequate to form any effective test of his views, his memoirs fell into oblivion" [Yule 1925, 4]. Apparently, some 80 years later, Pearl and Reed had arrived independently at the same result. Verhulst's work did eventually come to their attention; in fact, Yule acknowledges [Yule 1925, 5] that he is indebted to Pearl's book [Pearl 1922] for the references to Verhulst. Verhulst wrote in French; but with dictionary in hand, and a rudimentary high-school background in the language (like my own), the text is quite comprehensible.

This Module uses original data, diagrams, and text from these three original sources. The numbering of equations and figures follows that in the original, so it is not consistent throughout the Module. Also, one should be alert to changes in notation (population is represented as p or y, time as t or x, respectively). Some of the notation may also be confusing if the text is not read carefully; for example, p' and y' represent particular values of p and y, *not* derivatives. It is assumed that the audience understands and can work with geometric and arithmetic progressions and has had a basic introduction to differential equations. Quotations from original sources in English are either placed between quotation marks or else set off as displays with indented margins. Sources in French are rendered in split-page format, with the original French on the left and my translation on the right.

The exercises are designed to stimulate thought and inculcate the habit of reading mathematics with a pencil in hand, always ready to verify and check all claims made, and work out the equations for oneself.

This Module is intended to illustrate how mathematical knowledge grows—by fits and starts, rather than in a simple "linear" progression (as it is often presented in textbooks). Reading original sources, one notices that ideas are rediscovered and how later researchers borrow from and reinterpret the work of earlier mathematicians. Thus, in this Module, the same equations sometimes appear in slightly different forms, as they are reworked by various authors. The reader is encouraged to use these examples of what may at first appear to be redundancies in the text, as opportunities to compare and contrast different points of view, which can lead to further insights into the mathematics as well as its historical development.

2. The Logistic Equation

The logistic equation is used to model natural systems, involving growth with limited resources. This simple function, along with the differential equation that it satisfies and its familiar S-shaped curve, is ubiquitous and familiar to mathematicians and natural and social scientists alike. In the excerpts that follow, one can trace the early history of this model and gain insight into the assumptions on which it is based.

2.1 Yule's Summary of Malthus's Argument

We begin with Yule's summary of the history of attempts to model populations.

> Malthus, as will be well remembered by anyone who has ever read the *Essay on the Principle of Population,* reaches his conclusions by a *reductio ad absurdum* argument—the argument, to put it briefly, that if the population of a confined area increases without limit in geometric progression there will soon be millions without any food. [Yule 1925, 2]

Exercises

1. Look up Malthus's essay [1798], which has often been reprinted. Write a short summary of the key points.
2. Explain what is meant by a reductio ad absurdum argument.

> And Malthus seems almost to enjoy the depicting of horrors (or horrours, if one may use the earlier spelling, which in some odd way seems to add enormously to the effect) . . . . Malthus assumes that the population will double every 25 years, while the produce will be doubled in the first 25

years, but after that will only continue to increase in arithmetic progressions (*Essay* (1798) pp. 56-8). "And at the conclusion of the first century the population would be 112 millions, and the means of subsistence only equal to the support of 35 millions, which would leave a population of 77 millions totally unprovided for." It is a shocking picture, and it leaves our feelings so harrowed as to be capable of little further sympathy with the plight of the world, in which "in two centuries and a quarter the population would be to the means of subsistence as 512 to 10."

[Yule 1925, 3]

Exercises

3. If a population doubles every 25 years, and is 112 million at the end of one hundred years, what was the initial population?

4. Given the initial population above, with the same doubling time of 25 years, what will the population be in 225 years?

5. Using the fact that "the population is to the means of subsistence as 512 to 10," calculate how many millions can be supported after 225 years.

6. If the means of subsistence grows arithmetically, how many more people can be supported every 25 years? (Hint: use your knowledge of how many people can be supported after 225 years, and the fact that 35 million people are supported at the end of one hundred years. Also recall that the produce doubled in the first 25 years.)

7. Suppose two bacteria are placed in a Petri dish with a fixed amount of space. At the end of one minute, the number of bacteria has doubled (that is, there are now four bacteria in the dish). If there is exactly enough space in the dish for 1024 bacteria, how long before the space runs out, if

a) the number of bacteria increases in a geometric progression (this is called *exponential growth*);

b) the number of bacteria increases in an arithmetic progression (this is called *linear growth*)?

But there is another and more serious disadvantage attaching to such a mode of argument; it tells us very little. The only conclusion that can be drawn is that a population, confined to a specified area, does *not* increase in geometric progression. As to the true form of the law of increase, the argument gives us no information. [Yule 1925, 3]

2.2 Verhulst's Argument

Verhulst, a Belgian, published his papers in French. Below is the original text and a rough translation of some of his memoir. Yule refers to this memoir later in his address.

Au nombre des causes qui exercent une action constante sur l'accroissement de la population, nous placerons la fécondité propre a l'espèce humaine, la salubrité du pays, les mœurs de la nation que l'on considère, ses lois civiles et religieuses. Quant aux causes variable que l'on ne peut pas regarder comme les accidentelles, elles se résument généralement dans la difficulté de plus en plus grande que la population éprouve à se procurer des subsistances, lorsqu'elle est devenue assez nombreuse pour que toutes les bonnes terres se trouvent occupées.

Quand on ne tient pas compte de la difficulté dont nous venons de parler, il faut admettre qu'en vertu des causes constantes, la population doit croître en progression géométrique. En effet, si 1000 âmes sont devenue 2000 au bout de 25 ans, par exemple, il n'ya pas de raison pour que ces 2000 ne deviennent pas 4000 au bout de 25 années suivantes. [Verhulst 1845, 4]

The causes that exert a constant effect on the growth of population are: fertility, the wealth of the country, the death rate, and the nation's civil and religious laws. As for variable causes that aren't accidental, we must consider the difficulty in finding resources when the population becomes too numerous and all the good land is occupied.

If we consider only the constant causes, the population must grow in a geometric progression. In other words, if 1000 people become 2000 in 25 years, for example, there is no reason that 2000 should not become 4000 in the following 25 years.

Exercise

8. Does Verhulst's argument make sense to you? His is the first attempt to model population growth quantitatively. He tries to capture our common-sense notion of how populations grow, when such factors as birth and death rate are constant. Think about rabbits. If you start with 10 and the population doubles in 25 days, does it seem reasonable that if you started with 20, you would have 40 rabbits in 25 days? In what situations would this *not* be a reasonable assumption?

Les États Unis nous offrent un exemple de cette grande vitesse d'accroissement de la population. On y comptait, d'après le recensements officiels, [Verhulst 1845, 4]	The United States [in the late eighteenth and early nineteenth centuries] offers just such an example of a rapidly growing population that is expanding as if it had unlimited resources. A list of the official census figures follows.

En 1790	....	3,929,827	âmes [souls]
1800	....	5,305,925	
1810	....	7,239,814	
1820	....	9,638,151	
1830	....	12,866,020	
1840	....	17,062,566	

[Verhulst 1845, 4]

Si l'on prend pour la population de 1795 le chiffre 4,617,876, moyen entre celui de 1790 et celui de 1800, et qu'on fasse de même pour les années 1805, 1815, 1825 and 1835, on pourra évaluer approximativement les progrès de la population de 5 en 5 ans. C'est ainsi que nous avons formé le tableau suivant, dans lequel nous avons arrondi les chiffres et désigné par r le rapport de chaque population à celle qui la précède de 25 ans: [Verhulst 1845, 5]	In the following table, we take these official census figures for decades, and approximate the population in inter-censal years using the arithmetic mean. The third column lists the ratio, r, of each population to that of the preceding 25 years. The numbers are rounded.

This table (on the next page) illustrates a defining characteristic of exponential growth: for equal increments of time (in this case, 25-year intervals), the ratio between succeeding populations is constant (in this case about 2.1).

ANNÉES.	POPULATION.	VALEURS DE *r*.
1790.	3,930,000	
1795.	4,618,000	
1800.	5,306,000	
1805.	6,273,000	
1810.	7,240,000	
1815.	8,439,000	2.147
1820.	9,638,000	2.087
1825.	11,252,000	2.120
1830.	12,866,000	2.052
1835.	14,964,000	2.070
1840.	17,063,000	2.021

[Verhulst 1845, 5]

Nous n'insisterons pas davantage sur l'hypothèse de la progression géométrique, attendu qu'elle ne se réalise que dans des circonstances tout à fait exceptionnelles; par exemple, quand un territoire fertile et d'une étendue en quelque sorte illimitée, se trouve habité par un peuple d'une civilisation très-avancé, comme celle des premiers colons des États-Unis. [Verhulst 1845, 6]

[Unlike Malthus,] We readily admit that the hypothesis that populations increase in geometric progression is valid only in exceptional circumstances, as for example when a fertile and vast territory is inhabited by a technologically advanced people, like the early colonists in the United States.

2.3 Pearl and Reed's Data and Methods

Here is the opening paragraph of Pearl and Reed's paper.

It is obviously possible in any country or community of reasonable size to determine an empirical equation, by ordinary methods of curve fitting, which will describe the normal rate of population growth. Such a determination will not necessarily give any inkling whatever as to the underlying organic laws of population growth in a particular community. It will simply give a rather exact empirical statement of the nature of the changes which have occurred in the past. No process of empirically graduating raw data with a curve can in and of itself demonstrate the fundamental law which causes the occurring change. In spite of the fact that such mathematical expressions of population growth are purely empirical, they have a distinct and considerable usefulness. This usefulness arises out of the fact that actual counts of population by census methods are made at only relatively infrequent intervals, usually 10 years and practically never oftener than 5 years. For many statistical purposes, it is necessary to have as accurate an estimate as possible of the population in inter-censal years. This applies not only to the years following that on which the last census was taken, but also to the inter-censal years lying between prior censuses. For purposes of practical statistics it is highly important to have these inter-censal estimates of population as accurate as possible, particularly for the use of the vital statistician, who must have these figures for the calculation of annual death rates, birth rates and the like. [Pearl and Reed 1920, 275]

TABLE 1

SHOWING THE DATES OF THE TAKING OF THE CENSUS AND THE RECORDED POPULATIONS FROM 1790 TO 1910

DATE OF CENSUS		RECORDED POPULATION (REVISED FIGURES FROM STATISTICAL ABST., 1918)
Year	Month and Day	
1790	First Monday in August	3,929,214
1800	First Monday in August	5,308,483
1810	First Monday in August	7,239,881
1820	First Monday in August	9,638,453
1830	June 1	12,866,020
1840	June 1	17,069,453
1850	June 1	23,191,876
1860	June 1	31,443,321
1870	June 1	38,558,371
1880	June 1	50,155,783
1890	June 1	62,947,714
1900	June 1	75,994,575
1910	April 15	91,972,266

Table 1 from Pearl and Reed [1920, 277].

Exercise

9. Consider the data from Pearl and Reed's **Table 1**. After rounding up or down to the nearest thousand, estimate the population for the inter-censal years, assuming that the population is increasing from 1790 to 1910 in

a) a geometric progression;

b) an arithmetic progression.

Which estimate do you think is better? Why? How might you improve your estimate?

The usual method followed by census offices in determining the population in inter-censal years is one or the other of two sorts, namely, by arithmetic progression or geometric progression. These methods assume that for any given short period of time the population is increasing either in arithmetic or geometric ratio. Neither of these assumptions is ever absolutely accurate even for short intervals of time, and both are grossly inaccurate for the United States, at least, for any considerable period of time. What actually happens is that following any census estimates are made by one of another of these methods of the population for each year up to the next census, on the basis of data given by the last two censuses only. When that next census has been made, the previous estimates of the inter-censal years are corrected and adjusted on the basis of the facts brought out at that census period. [Pearl and Reed 1920, 275–276]

Exercises

10. Given the data in **Table 1**, how would you determine an empirical equation that fits the data? Do not actually find such an equation, just explain how you would go about it.

11. What is the difference between finding an equation of "best fit" for a given set of data, and determining a "fundamental law" that "causes the occurring change"?[1]

We continue the quotation from Pearl and Reed's paper:

It would be the height of presumption to attempt to predict *accurately* the population a thousand years hence. But any real law of population growth ought to give some general and approximate indication of the number of people who would be living at that time within the present area of the United States, provided no cataclysmic alteration of circumstances has in the meantime intervened.

It has seemed worth while to attempt to develop such a law, first by formulating a hypothesis which rigorously meets the logical requirements, and then by seeing whether in fact the hypothesis fits the known facts. The general biological hypothesis which we shall here test embodies as an essential feature the idea that the rate of population increase in a limited area at any instant of time is proportional (a) to the magnitude of the

[1]Instructors may want to discuss the work of Brahe, Kepler, and Newton in this context, or ask students to research this. Brahe was an observer who collected the most accurate data of his time on the motions of the planets. Kepler discerned patterns in the data and derived equations to describe the paths planets followed (ellipses) and relationships between a planet's period of revolution and its distance from the sun. Newton explained the observations through a general law (of gravitation) that implied Kepler's equations and much more. Kepler used simple induction to express a regularity of nature, while Newton may be said to have discovered a fundamental causal relationship. See, for example, Kuhn [1970, 209–219] and Abers and Kennel [1977, 105–132].

population existing at that instant (amount of increase already attained) and (b) to the still unutilized potentialities of population support existing in the limited area. [Pearl and Reed 1920, 281]

Exercise

12. Let y represent the population at time x. Write an equation for the relationship between dy/dx (the rate of population increase) and the population that models the above hypotheses.

The following conditions should be fulfilled by any equation which is to describe adequately the growth of population in an area of fixed limits.

1. Asymptotic to a line $y = k$ when $x = +\infty$.
2. Asymptotic to a line $y = 0$ when $x = -\infty$.
3. A point of inflection at some point $x = \alpha$ and $y = \beta$.
4. Concave upwards to left of $x = \alpha$ and concave downward to right of $x = \alpha$.
5. No horizontal slope except at $x = \pm\infty$.
6. Values of y varying continuously from 0 to k as x varies from $-\infty$ to $+\infty$.

In these expressions y denotes population, and x denotes time. [Pearl and Reed 1920, 281]

Exercise

13. Give reasons why "any equation which is to describe adequately the growth of a population in an area of fixed limits" should satisfy each of the six conditions listed.

a) Draw a graph that illustrates each of the conditions separately.

b) Draw one graph that meets all of the conditions simultaneously.

An equation which fulfills these requirements is

$$y = \frac{be^{ax}}{1 + ce^{ax}} \tag{ix}$$

when a, b and c have positive values. [Pearl and Reed 1920, 281]

Exercise

14. Verify that equation **(ix)** of Pearl and Reed meets each of the conditions **(1–6)**.

In this equation the following relations hold:

$$x = +\infty \qquad y = b/c \tag{x}$$

$$x = -\infty \qquad y = 0 \tag{xi}$$

Relations (x) and (xi) define the asymptotes. The point of inflection is given by $1 - ce^{ax} = 0$, or

$$x = -(1/a)\log c \qquad y = b/2c \tag{xii}$$

The slope at the point of inflection is $ab/4c$.

[Pearl and Reed 1920, 281–282]

Exercise

15. Verify the relations in **(x)**, **(xi)**, and **(xii)**.

Expressing the first derivative of (ix) in terms of y, we have

$$\frac{dy}{dx} = \frac{ay(b - cy)}{b} \tag{xiii}$$

[Pearl and Reed 1920, 282]

Exercise

16. Compare this equation with the one that you came up with above, in **Exercise 12**, to model Pearl and Reed's hypotheses. Show that if one lets $L = b/c$, then the equation above can be written in terms of only two constants, a and L. The third constant in **(ix)** arises as a constant of integration and depends on initial conditions.

The general form of the curve (ix) is shown in figure 2.

Putting the equation in this form shows at once that it is identical with that describing an autocatalyzed chemical reaction, a point to which we shall return later.

[Pearl and Reed 1920, 282]

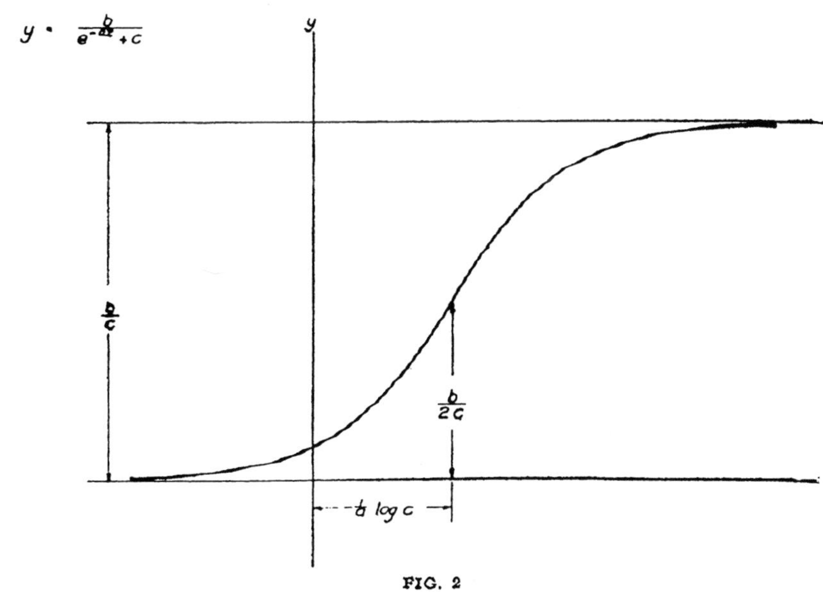

FIG. 2
General form of curve given by equation (ix).

Figure 2 from Pearl and Reed [1920, 282].

Exercise

17. Look up the definition of *autocatalysis*. In what ways is this process similar to population growth in an area of limited resources?

> There is much that appeals to the reason in the hypothesis that growth of population is fundamentally a phenomenon like autocatalysis. In a new and thinly populated country the population already existing there, being impressed with the apparently boundless opportunities, tends to reproduce freely, to urge friends to come from older countries, and by the example of their well-being, actual or potential, to induce strangers to immigrate. As the population becomes more dense and passes into a phase where the still unutilized potentialties of subsistence, measured in terms of population, are measurably smaller than those which have already been utilized, all of these forces tending to the increase of population will become reduced. [Pearl and Reed 1920, 287]

2.4 Continuation of Yule's Account

We now return to Yule's historical account.

Verhulst, Professor of Mathematics at the École Militaire, . . . states [in a memoir from 1838] that he had long since attempted to determine the probable form of the law of population, but had abandoned the investigation on account of the inadequacy of the data. But, as the course he had followed would, as it seemed to him, necessarily lead to the true law when sufficient data were available, and as the results at which he had arrived were of some interest, he had consented to M. Quetelet's invitation to publish them. Let p denote the population, t the time; then if the population is increasing in geometric progression

$$dp/dt = mp.$$

[Yule 1925, 43]

Exercise

18. What is the solution to this differential equation? Does this clarify the connection between "growing exponentially" and "growing in a geometric progression"? Explain.

But since the rate of growth of the population is retarded by the increased number of the inhabitants, we must subtract from mp some unknown function of p, so that the differential equation to be integrated takes the form

$$dp/dt = mp - \varphi(p).$$

The simplest assumption that can be made as to the form of $\varphi(p)$ is to suppose $\varphi(p) = np^2$, which gives as the solution

$$p = \frac{mp'e^{mt}}{np'e^{mt} + m - np'}. \qquad (*)$$

where p' is the population at zero time, and the limiting population when t is infinite is m/n. [Yule 1925, 43]

Exercises

19. Verify that the function given for p above in **(*)** does satisfy the differential equation $dp/dt = mp - np^2$. Then show that as $t \to \infty$, $p \to m/n$.

20. Explain how $\varphi(p) = np^2$ is the "simplest" assumption that can be made about the form of $\varphi(p)$.

Verhulst returns to the subject in a much longer memoir a few years later. [This is the memoir quoted in this Module.] The argument is here developed on slightly different and simpler lines. The freely-expanding population, it is admitted, must increase in geometric progression, the data for the U.S.A. 1790–1840 being used to illustrate the point. But suppose that the population has expanded up to the point when "the difficulty of finding good land has begun to make itself felt." Let the population at this epoch, which will be taken as zero time, be b: b is termed by Verhulst the "normal population." The "retarding function" now comes into play, and the differential equation may be written

$$\frac{1}{p}\frac{dp}{dt} = l - f(p-b).$$

(The retarding function is now, more naturally, taken as a retarding function for the logarithmic differential instead of dp/dt.)

[Yule 1925, 43–44]

Exercises

21. What is a logarithmic differential?

22. Explain, in your own words, the role of the "retarding function" in this model of population growth.

Only two conditions are necessary for the retarding function in its new form: it must increase indefinitely with the population, and it must vanish when $p = b$. [Yule 1925, 44]

Exercise

23. Justify and explain why these two conditions are necessary for the "retarding" function.

The simplest form to assume is $n(p-b)$: we then have: – –

$$\frac{1}{p}\frac{dp}{dt} = l - n(p-b),$$

or, writing for brevity $m = l + nb$,

$$\frac{1}{p}\frac{dp}{dt} = m - np.$$

Verhulst now christens the curve a "logistic." He develops the principal properties, pointing out that the curve is symmetrical with respect to the point of inflection, and that the ordinate at the point of inflection is half the limiting ordinate. [Yule 1925, 44]

2.5 Return to Verhulst's Original Account

Here is the relevant passage from Verhulst.

Désignons par p la population, par t le temps, et par k et l des constantes indéterminées: si la population croît en progression géométrique pendant que le temps croît en progression arithmétique, on aura entre ces deux quantités la relation,

Let p be the population and t stand for time, with k and l undetermined constants. If the population grows in a geometric progression, while time grows in arithmetic progression, the two quantities will be related in the following way:

$$p = k \cdot 10^{lt}.$$

[Verhulst 1845, 5]

Exercise

24. Compare this to the solution that you got in **Exercise 18** above.

Soit p' une population correspondante à un temps t': il viendra

If p' is the population corresponding to time t', then this becomes

$$p = p' \cdot 10^{l(t-t')},$$

et si l'on appelle π la population existante au moment d'où l'on commence à compter le temps, l'équation précédente devient

and if one lets π be the population at the time one starts counting, the preceding equation becomes

$$p = \pi \cdot 10^{lt}. \quad \textbf{(1)}$$

... La période malthusienne de 25 ans suppose que p devient $2p$ quant t devient $t+25$, l'année étant prise pour unité de temps: on a donc le équation (...):

The "malthusian period" of 25 years assumes that p becomes $2p$ when t becomes $t + 25$, the year being taken as the unit of time. One then finds that

$$l = (1/25) \log 2 = .012041200.$$

[Verhulst 1845, 5–6]

Exercise

25. Verify each of the above equations. (Note that here log denotes a logarithm to the base 10. We will use ln to denote natural logarithms.)

La différentiation de l'équation (1) donne	Differentiating equation (1) gives

$$\frac{M}{p}\frac{dp}{dt} = l, \tag{2}$$

... et en désignant par M le module par lequel il faut multiplier les logarithmes népériens pour les convertir en logarithmes vulgaires. [Verhulst 1845, 6]	where $M = \log e$.

Exercise

26. Show that if $M \ln x = \log x$, then $M = \log e$, for any real number x.

Cette quantité étant constante, on peut la prendre pour mesure de l'énergie avec laquelle la population tend à se développer, lorsqu'elle n'est point retenue par la crainte de manquer de subsistances. On a aussi, avec une exactitude d'autant plus grande que Δp et Δt sont plus petits,	The ratio of the rate of change of the population to the population itself is thus constant, and one can take this constant to be a measure of the energy with which the population tends to grow, when not constrained by limited resources. In fact, for small changes in p and t (Δp and Δt), we may say

$$M\Delta p = lp\Delta t;$$

et, si l'on prend pour Δt l'intervalle d'une année,	and if one takes Δt to be one year, we arrive at

$$\frac{\Delta p}{p} = \frac{l}{M},$$

c'est-à-dire que, dans le cas de la progression géométrique, l'excès annuel des naissances sur les décès, divisé par la population qui l'a fourni, donne un quotient constant .... C'est un fait d'observation que, dans toute l'Europe, le rapport de l'excès annuel des naissances sure les décès à la population qui l'a fourni, et par conséquent le coefficient l/M, va sans cesse en s'affaiblissant: de manière que l'accroissement annuel, dont la valeur absolue augmente continuellement lorsqu'il y a progression géométrique, paraît suivre un progression tout au plus arithmétique. Cette remarque confirme le célèbre aphorisme de Malthus, que *la population tend à croître en progression géométrique, tandis que la production des subsistances suit une progression tout au plus arithmétique*. [Verhulst 1845, 7]

which is to say that in the case of the population growing in geometric progression, the excess of annual births over deaths, divided by the population, is a constant ratio. However, throughout Europe, it is observed that this ratio, l/M, in fact decreases. This observation confirms the celebrated aphorism of Malthus, that *the population tends to grow in geometric progression while the production of food follows a more or less arithmetic progression*.

As you read Verhulst's original derivation below of the logarithmic differential equation, compare it to Yule's treatment above (on p. 13, following **Exercise 23**).

On peut faire un infinité d'hypothèses sur la loi d'affaiblissement du coefficient l/M. La plus simple consiste à regarder cet affaiblissement comme proportionnel à l'accroissement de la population, depuis le moment où la difficulté de trouver de bonnes terres a commencé à se faire sentir. Nous appellerons *population normale*, et nous désignerons par b, celle qui correspond à cette époque remarquable, à partir de laquelle nous compterons le temps: puis, ayant dénoté par n un coefficient indéterminé, nous remplacerons l'équation différentielle

One could make an infinite number of hypotheses about the law of decrease of the coefficient l/M. The simplest is to consider the decrease to be proportional to the growth of the population, from the time when the difficulty of finding good land begins to be felt. We will begin counting from this time, and call the population at this time the *normal population*, designated by b. Then, letting n denote an undetermined coefficient, we replace the differential equation

$$\frac{M}{p}\frac{dp}{dt} = l,$$

relative à la progression géométrique, par	by

$$\frac{M}{p}\frac{dp}{dt} = l - n(p - b),$$

d'où, en posant, pour abréger, $m = l + nb$,	and substituting $m = l + nb$,

$$\frac{M}{p}\frac{dp}{dt} = m - np,$$

et	and

$$dt = \frac{M\,dp}{mp - np^2}.$$

Cette équation étant intégrée donne, en observant que $t = 0$ répond à $p = b$,	We integrate this equation, noticing that $t = 0$ corresponds to $p = b$,

$$t = \frac{1}{m}\log\frac{p(m - nb)}{b(m - np)}.$$

Nous donnerons le nom de *logistique* à la courbe caractérisée par l'équation précédente. [Verhulst 1845, 8–9]	and give the name *logistic* to the curve characterized by the previous equation.

How exciting! Here is where Verhulst first "christens" the equation a logistic. Why? In its modern incarnation, the logistic equation is usually written with population expressed as a function of time (population as the dependent variable). This perhaps more familiar form of the equation involves an exponential. Verhulst wrote the relationship here with time as the dependent variable. Since the log function is the inverse of the exponential, his equation has t (time) equal to a (somewhat complicated) logarithmic function of p (population). Thus "logistic" is meant to convey the curve's "log-like" quality. For further discussion of this point, see Shulman [1997].

Exercises

27. Perform the integration indicated, and verify the equation for t. Why doesn't M appear in the expression for t? (Recall that $M = \log e$.)

28. Graph t as a function of p.

29. Recall Yule's statement: "Verhulst now christens the curve a 'logistic.' He develops the principal properties, pointing out that the curve is symmetrical with respect to the point of inflection, and that the ordinate at the point of inflection is half the limiting ordinate." Verify that the curve is symmetrical with respect to the point of inflection, and that the ordinate at the point of inflection is half the limiting ordinate.

30. Express p as a function of t. Do you expect this curve to have the same properties? Graph p as a function of t.

2.6 Further History and Yule's Own Development

Yule continues his story.

But the work of Verhulst, as I have said, was forgotten. Only some four years ago, Professors Pearl and Reed of the Johns Hopkins University, Baltimore, working on interpolation formulae for population with especial reference to the United States, arrived independently at precisely the same result. After trying sundry purely empirical formulae, they point out that no such formula can be regarded as a general law of population growth, however good it may prove for practical purposes over a limited period. General considerations suggest something as to the form of the rational law. As there must be some limit to the population on the given area, the curve of growth must, sooner or later, turn over (i.e., in mathematical terms pass through a point of inflection) and gradually approach that limit. If we assume that the *absolute* rate of growth of the population (i.e., the numbers added to the population per unit of time, not the percentage rate of increase) is proportional to (1) the magnitude of the population existing at that instant, (2) "the still unutilized reserves of population-support existing" in the confined area, or in other words the differences between the existing and the limiting population, we arrive at precisely the form of law suggested by general considerations, and the formula is that given by Verhulst. Pearl and Reed's discovery was, however, quite independent, and their work on this subject seems to me of the highest importance and interest for the theory of population . . . .

I prefer to write Verhulst's formula for the law of growth in the form

$$y = \frac{L}{1 + e^{(\beta - t)/\alpha}}. \tag{1}$$

where y is the population, t the time and L the limiting value of the population, which is only approached as t becomes indefinitely great.

[Yule 1925, 4–5]

Exercise

31. Compare this with the equation that you obtained in **Exercise 29** above.

There are two other constants in addition to L, viz., α and β. Of these, α determines the horizontal scale of the curve—the greater α the more the curve spreads out—and I propose to call it the *standard interval*: β is the time from the zero of the time-scale to the point of inflection. Not to make the text of my Address too technical, I have relegated to Appendix II some discussion of the mathematics of the curve, which, following Verhulst, we may term a "logistic." Here it is only necessary to direct attention to some of its principal properties. If we choose the point of inflection as zero time, the standard interval as our unit of time, and the limiting population L as the unit of population, the formula (1) takes its simplest form

$$y' = \frac{1}{1+e^{-\tau}}.$$

Fig. 1 [on the following page] shows the curve drawn from this formula. It starts rising very slowly and near the base line, gradually turns up more and more steeply till it reaches the point of inflection, and then gets flatter and flatter as it approaches the limit. It is symmetrical round the point of inflection, in the sense that if y' and y'' are ordinates of the curve equidistant from the point of inflection to left and right

$$y' = 1 - y''.$$

This is clearly a limitation on the generality of the curve, but only experience can tell us how far the symmetrical form is likely to be valid: both Verhulst and Pearl and Reed discuss more general forms. The proportional rate of increase at any instant of time, in the curve drawn as in fig. 1, is given by the complement of the ordinate, i.e., the intercept between the curve and the horizontal line representing the limiting population. It is obvious from the figure that at first, when the population is still very small, the proportional rate of increase only changes very slowly, so that the growth of the population can hardly be distinguished from growth in geometric progression; but as time goes on it falls more and more rapidly until the point of inflection is passed. It is important to note that in such a curve the proportional (or percentage) rate of increase of the population *falls continuously* from the start; if the percentage rate of increase of a population is steadily *rising* (mere disturbances excluded) it cannot be regarded as following a simple logistic cycle. It may be that such a population is passing from a cycle with a longer standard-interval to a cycle with a shorter standard-interval, e.g., when an agricultural country starts developing industries: or it may be that the population should be regarded as a mixture or association of two distinct populations following separate cycles.

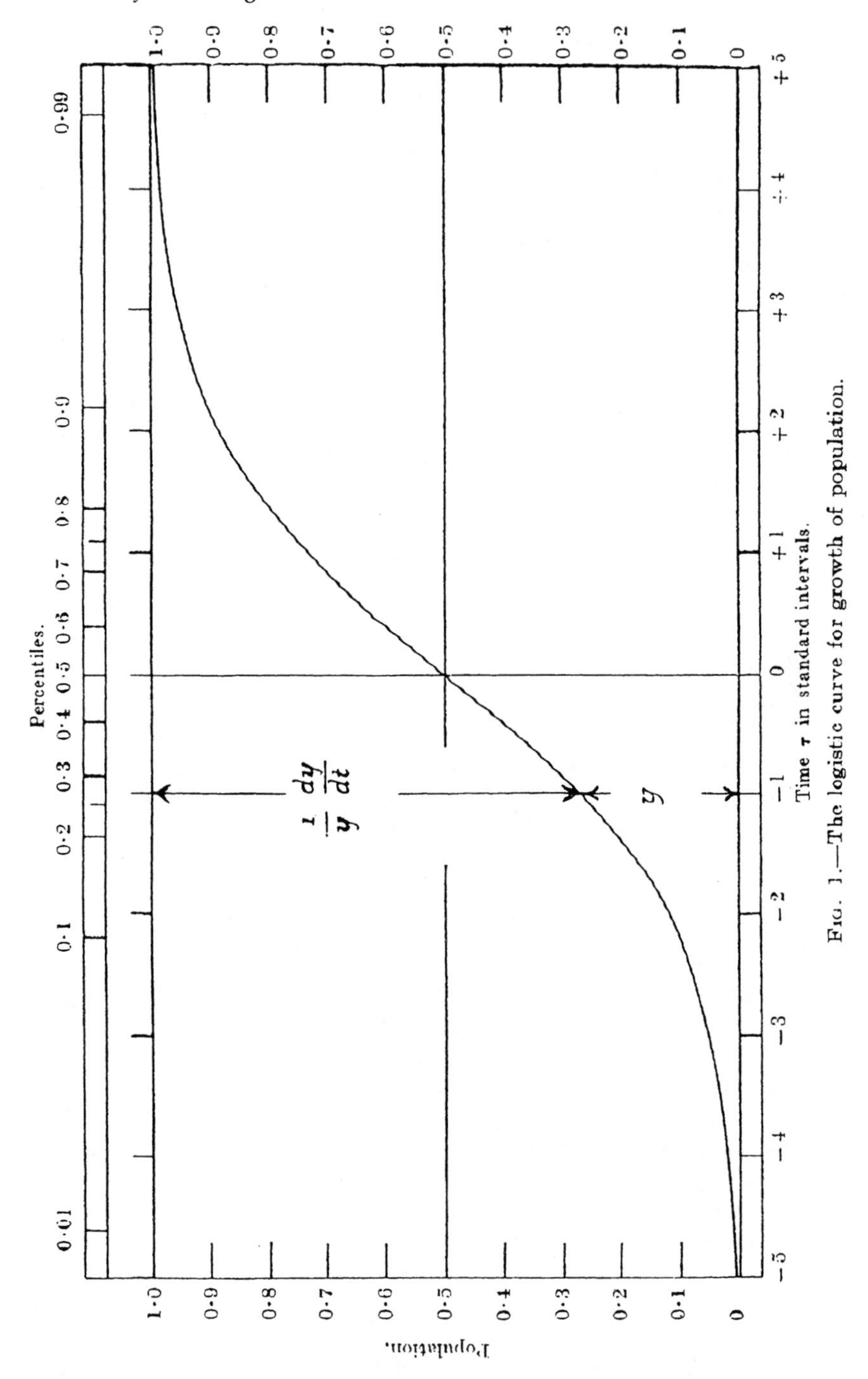

FIG. 1.—The logistic curve for growth of population.

Figure 1 from Yule [1925, 6]. ©Royal Statistical Society. Reproduced with permission.

From the scale of percentiles at the top of fig. 1 it will be seen that the population stands at only 1 per cent of its limiting values at -4.6τ and reaches 99 percent of the limiting value at $+4.6\tau$. It therefore passes through the great bulk of the cycle in 9 or 10 standard intervals. The quartile and decile points lie at 1.1 and 2.2 respectively, very nearly indeed. The central 80 per cent of the range from 0 to L is therefore covered in roundly 4.4 standard intervals, and the central half in only 2.2 intervals.

[Yule 1925, 5–7]

2.7 Yule's Appendix

APPENDIX II. -- *Some notes on the mathematics of the logistic curve and methods of fitting*[2]. Let the differential equation be written: --

$$(1/y)dy/dt = (1/\alpha)(1 - (y/L)) \tag{1}$$

In this form of the equation L is evidently the limiting population, since dy/dt is zero when $y = L$, and the constant α must be of the dimensions of a time. I shall term it the "standard interval," by analogy with the "standard deviation." The solution of the differential equation is

$$y = \frac{L}{1 + e^{(\beta - t)/\alpha}}.$$

where β is a constant of integration, and is evidently also a time. When t is infinite $y = L$: when $t = \beta$, $y = L/2$. But, differentiating (1) again,

$$d^2y/dt^2 = (1/\alpha)(1 - (2y/L)),$$

and hence $y = L/2$, $t = \beta$, gives the point of inflection. Further,

$$\begin{aligned} y_{\beta+h} &= L/(1 + e^{-h/a}) = L - L/(1 + e^{h/a}) \\ &= L - y_{\beta+h}. \end{aligned}$$

Hence the curve is symmetrical about the point of inflection.

The smaller y is compared with L the more nearly does the differential equation approach the simple form

$$(1/y)(dy/dt) = 1/\alpha.$$

But the solution of this is a logarithmic [we would say exponential] curve

$$y = Ae^{t/\alpha}.$$

That is to say, the early stages of the logistic are sensibly the same as a logarithmic [exponential] curve, or the curve of a geometric progression.

[2]For a more modern approach, see, e.g., Cavallini [1993].

We would then, in any case, expect the early stages of the growth of a population to be appreciably geometric; there does not seem to be any necessity for Verhulst's conception of an initial stage in which the growth is strictly geometric, passing abruptly into the logistic when the "normal population" is reached.

If we measure time with the standard interval as unit, denoting the time so measured by τ, take the point of inflection as the origin for time, and measure population with the limiting population L as unit, writing y' for y/L, we have

$$y' = \frac{1}{1+e^{-\tau}}. \tag{2}$$

This is the simplest form of the equation to the logistic, and its differential equation is

$$(1/y')(dy'/dt) = 1 - y'.$$

Evidently we could draw such a logistic once and for all and fit the data for any actual population thereto by (1) replacing the actual populations by their ratios to the limiting population, (2) making the points of inflection coincide, (3) taking the standard time as the unit of our time-scale.

[Yule 1925, 46–47]

Exercise

32. Take three sets of data on various populations, "normalize" them as Yule suggests above, and then fit them to the normalized curve **(2)**.

3. Moral

The moral of this story[3] is:

The fundamental property of the logistic is that the instantaneous percentage rate of increase is a linear function of the population (equation (1)).

[Yule 1925, 48]

---

[3]But the story does not end here. The logistic model continues to yield new and exciting mathematics. For instance, the discrete-time version of the logistic population model can lead to chaotic dynamics. See, e.g., Schroeder [1991, 268ff].

4. Solutions to Selected Exercises

3.

Year	0	25	50	75	100	...	$25n$
Population	P_0	$2P_0$	$4P_0$	$8P_0$	$16P_0$	...	$2^n P_0$

Since $16P_0 = 112$, $P_0 = 7$ million.

4. In 225 years, the population doubles nine times; $7 \times 2^9 = 3{,}584$ million.

5. Let x be the number of millions supported. Then, since

$$\frac{7 \times 2^9}{x} = \frac{512}{10} = \frac{2^9}{10},$$

we have $x = 70$; so the answer is 70 million.

6.

Year	...	100	125	...	225
People supported	...	35	$35 + x$	...	$35 + 5x$

We have $35 + 5x = 70$, so $x = 7$. So the answer is 7 million every 25 years.

Alternatively, consider that we start by supporting $P_0 = 7$ million people. The number doubles in 25 years, so we have $14 = 7 + x$, or $x = 7$; and we get the same answer as before, 7 million every 25 years.

7. a)

Minute	0	1	2	3	...	n
Number of bacteria	2	4	8	16	...	2^{n+1}

We have $2^{n+1} = 2^{10}$, hence $n + 1 = 10$ and $n = 9$ min.

b)

Minute	0	1	2	3	...	n
Number of bacteria	2	4	6	8	...	$2(n + 1)$

We have $2(n + 1) = 2^{10}$, hence $n + 1 = 2^9$ and $n = 2^9 - 1 = 511$ min.

9. a) The population grows by a factor of $\left(\frac{91{,}972}{3{,}929}\right)^{1/12} = 1.3005$ each decade, or by a factor of $(1.3005)^{1/10} = 1.0266$ each year.

b) $(91{,}972 - 3{,}929)/120 = 734$ thousand/year.

12. $dy/dx = ay(b - y)$.

18. $p = Ke^{mt}$.

30. $p = \dfrac{mbe^{mt/M}}{m - nb + nbe^{mt/M}}$.

References

Abers, Ernest, and Charles Kennel. 1977. *Matter in Motion: The Spirit and Evolution of Physics.* Boston, MA: Allyn and Bacon, Inc.

Cavallini, Fabio. 1993. Fitting a logistic curve to data. *College Mathematics Journal* 24: 247–253.

Kuhn, Thomas. 1970. *The Copernican Revolution.* Cambridge, MA: Harvard University Press.

Malthus, T.R. 1798. *An essay on the principle of population, etc.* London: Anon.

Pearl, Raymond. 1922. *The Biology of Death.* London, England: J.B. Lippincott.

________, and Lowell Reed. 1920. On the rate of growth of the population of the United States since 1790 and its mathematical representation. *Proceedings of the National Academy of Sciences* 6 (6) (15 June 1920): 275–288.

Schroeder, Manfred. 1991. *Fractals, Chaos and Power Laws: Minutes from an Infinite Paradise.* New York: W.H. Freeman.

Shulman, Bonnie. 1997. MATH-ALIVE!: Using Original Sources to Teach Mathematics in Social Context. *PRIMUS: Problems, Resources and Issues in Undergraduate Studies*, forthcoming.

Verhulst, Pierre-François. 1845. Recherches mathématiques sur la loi d'accroissement de la population. *Nouveaux Mémoires de l'Académie Royale des Sciences et Belles-Lettres de Bruxelles* 18: 1–38.

Yule, G. Udny. 1925. The growth of population and the factors which control it. *Journal of the Royal Statistical Society* 88, Part 1 (January 1925): 1–90.

About the Author

Bonnie Shulman received a Ph.D. in mathematical physics from the University of Colorado at Boulder in 1991. Since then she has been Assistant Professor of Mathematics at Bates College in Lewiston, ME. Her research interests include tracing the historical development of mathematical concepts, which she then incorporates in her teaching. She hopes that her students enjoy it as much as she does!

UMAP

Modules in Undergraduate Mathematics and Its Applications

Published in cooperation with

The Society for Industrial and Applied Mathematics,

The Mathematical Association of America,

The National Council of Teachers of Mathematics,

The American Mathematical Association of Two-Year Colleges,

The Institute for Operations Research and the Management Sciences, and

The American Statistical Association.

Module 767

The Mathematics of Scuba Diving

D. R. Westbrook

Applications of Beginning Calculus to Physiology

COMAP, Inc., Suite 210, 57 Bedford Street, Lexington, MA 02173 (781) 862–7878

INTERMODULAR DESCRIPTION SHEET: UMAP Unit 767

TITLE: The Mathematics of Scuba Diving

AUTHOR: D.R. Westbrook
Dept. of Mathematics and Statistics
University of Calgary
Calgary, Alberta, Canada T2N 1N4
westbroo@@acs.ucalgary.ca

MATHEMATICAL FIELD: Beginning calculus

APPLICATION FIELD: Physiology

TARGET AUDIENCE: Students in beginning calculus

ABSTRACT: Exponential solutions of differential equations are used to construct decompression schedules for dives of various durations to various depths.

PREREQUISITES: A knowledge of differential and integral calculus related to exponential functions.

RELATED UNITS: Unit 676: *Compartment Models in Biology*, by Ron Barnes. *The UMAP Journal* 8 (2): 133–160. Reprinted in *UMAP Modules: Tools for Teaching 1987*, edited by Paul J. Campbell, 207–234. Arlington, MA: COMAP, 1988.

Tools for Teaching 1997, 191–219.

COMAP, Inc., Suite 210, 57 Bedford Street, Lexington, MA 02173
(800) 77-COMAP = (800) 772-6627, or (781) 862-7878; http://www.comap.com

The Mathematics of Scuba Diving

D.R. Westbrook
Dept. of Mathematics and Statistics
University of Calgary
Calgary, Alberta, Canada T2N 1N4
westbroo@@acs.ucalgary.ca

Table of Contents

Modules and Monographs in Undergraduate Mathematics and its Applications (UMAP) Project

The goal of UMAP is to develop, through a community of users and developers, a system of instructional modules in undergraduate mathematics and its applications, to be used to supplement existing courses and from which complete courses may eventually be built.

The Project was guided by a National Advisory Board of mathematicians, scientists, and educators. UMAP was funded by a grant from the National Science Foundation and now is supported by the Consortium for Mathematics and Its Applications (COMAP), Inc., a nonprofit corporation engaged in research and development in mathematics education.

Paul J. Campbell	Editor
Solomon Garfunkel	Executive Director, COMAP

1. Introduction

Are you a scuba diver? Can you use the diving tables? Do you know the mathematical basis for the diving tables? Could you construct your own diving tables? The purpose of this module is to describe the physiological basis for the diving tables and the mathematics used for the calculations.

2. A Brief History of Diving

Diving is an ancient pastime. Diving for profit—the collectinn of sponges, shells, and pearls—and diving for food have been with us for some time, and probably so has diving for pleasure. Divers were used for military purposes by the Greeks and are still of strategic importance today.

Ancient diving was essentially free (or breath-hold) diving, although Alexander the Great was reported to have used a primitive diving bell around 330 B.C. A diving bell is essentially a weighted inverted receptacle that retains its air (or other gases) as it is lowered into the water, giving a source of oxygen at depth to which the diver may return as needed or even be connected by a flexible tube. The air in the bell deteriorates in quality as the dive progresses, and various methods have been devised to replenish it.

In 1691, Sir Edmund Halley (of comet fame) built and patented what may have been the first practical diving bell, with a volume of approximately 60 cubic feet. The air was replenished from barrels, and the fouled air was vented out by means of a valve. (A 6-foot-high cylinder of diameter $3\frac{1}{2}$ ft has volume $\simeq$ 56 ft^3.) Nearly 100 years passed before a successful forcing pump was developed to enable a supply of fresh air to be pumped to the bell from the surface. This technique later developed into personal diving suits supplied from the surface and then to **s**elf-**c**ontained **u**nderwater **b**reathing **a**paratus (SCUBA).

As dives became deeper and longer, it became apparent that there were various physiological risks involved. One such risk is decompression sickness, or the "bends," which was associated with a rapid return to the surface after a long or deep dive.

In addition to diving, the nineteenth century saw the introduction of "caissons," large chambers equipped with an air lock and kept under high pressure, which enabled tunnellers and bridge builders to work underground or underwater without the chamber flooding. It soon became clear that special procedures were needed so that the workers, who may have been working in a high-pressure environment for several hours, did not suffer injuries or even death when they returned to normal atmospheric pressure. The need for a careful decompression sequence became obvious. In 1854, physicians B. Pol and T.J.J. Wattelle stated in a report, "The danger does not lie in entering a shaft containing compressed air; nor in remaining there a longer or shorter time; decompression alone is dangerous" [Hills 1977].

The decompression routines of this time were usually linear (i.e., a reduction in pressure at a fixed constant rate in atmospheres per minute) and were generally devised by experience that involved much pain and some deaths on the part of the experimental subjects. Of the approximately 600 men who worked on the St. Louis bridge, 119 suffered serious neurological decompression sickness, and 14 died. The name "the bends" apparently originated from the gait of these bridge workers, caused by pains in their joints. This resembled the "Grecian bend" of fashionable ladies of the time, who walked voluntarily in this manner.

In the early twentieth century, military needs led various navies to become interested in decompression sickness, and more careful research was begun. The most influential of this research was performed by the physiologist J.S. Haldane for the Royal Navy in 1906. Haldane's diving tables (1908) were remarkably effective in almost eliminating decompression sickness as a diving hazard and were used for some time. As more experience was gained, it became clear that Haldane's tables were somewhat conservative for short dives, so adjustments were made. Then, as longer deeper dives were undertaken, it was found that the tables were not conservative enough for such dives, and more refinements were made. Many further refinements have taken place in more recent times, but the tables are still essentially based on adaptations of Haldane's original ideas.

In the following sections, we examine these basic ideas and the mathematics behind them. To construct adequate universal tables is arithmetically intensive, but we will use the ideas in simplified form to construct our own tables.

The tables that we construct are not to be used in any dive!
Use the tables that your scuba instructors give you.

3. Haldane's Model

When Haldane began his experiments, it had been established that the major cause of decompression sickness was the release of bubbles of nitrogen, an inert gas in the air, into various tissues and into the arterial bloodstream. While a diver is underwater, she is breathing air under high pressure and, as a result, more nitrogen is forced into her blood. When she ascends, the air that she is breathing returns to a lower pressure, and the nitrogen dissolved in her blood forms bubbles. (Because oxygen in the air that is dissolved in the blood is metabolized, it does not cause a problem.) The effect can be seen when the lid of a pop bottle is unscrewed. The gas in the fluid is under pressure that is suddenly reduced when the lid is unscrewed, and bubbles rapidly form.

Initially it was thought that there would be a critical drop in pressure above which sickness would occur; but Haldane's experiments, which were performed on goats, led to a different conclusion. (Haldane had found that the sensitivity of goats to decompression sickness was acceptably close to that of humans.) He found that no matter what the original pressure is, decompression sickness does not occur if the pressure is reduced by less than some fixed fraction. That is, there is a value M for which a pressure P_1 can be reduced to $P_2 = MP_1$ without the occurrence of "the bends." Haldane suggested a value M just slightly less than 1/2. We will use $1/2.15 \approx .465$ in our calculations.

The subjects of these experiments were exposed to the higher pressure for long periods, so the dissolved gases were brought to saturation levels. In dives, this might not be the case. In addition, for long dives at an absolute pressure of more than twice atmospheric pressure, the subject could not be brought to atmospheric pressure without one or several intermediate stops. (An absolute pressure of two atmospheres occurs at a depth of about 10 m $\approx$ 33 ft of water.)

To determine an appropriate set of stops, a model of how gases are dissolved in and released from body tissues is needed. First, it is known that the pressure of inert gas in the pulmonary circuit is almost instantaneously equalized with that in the lungs, which is the ambient external pressure. Thus, blood entering the arterial system has gas pressure equal to the ambient pressure. A model must now be made of the distribution of the gas to the various tissues in the body.

The simple model that we use in this Module is based on the following assumptions:

- The blood flows through a tissue at a constant volume rate ν ml/sec.
- If the gas pressure in the blood and tissue is p, then the concentration of the gas in the blood is $s_1 p$ g/ml and in the tissue is $s_2 p$ g/ml, where s_1, s_2 are constants with different values of s_2 for different tissues.

The model is a simple compartment model (see Barnes [1987]). Gas enters the pulmonary circuit from the lungs at pressure p_e, the ambient external pressure. We assume that the gas pressure in the blood as it enters a tissue compartment is p_e. The pressure in the tissue and the blood is quickly equalized to the local pressure p, and the blood leaves the compartment at pressure p.

A balance of mass for the gas must hold:

The rate of increase of mass in the compartment =
Rate at which mass flows in − Rate at which mass flows out.

The mass of gas in the compartment at any time is $V_1 s_1 p + V_2 s_2 p$, where V_1 and V_2 are measured in ml and represent the respective volumes of blood and tissue in the compartment. The rate of increase of mass is then

$$\frac{d}{dt}[(V_1 s_1 + V_2 s_2)p] \quad \text{g/sec.}$$

Gas enters the compartment at a rate $\nu s_1 p_e$ g/sec and leaves at a rate of $\nu s_1 p$ g/sec. The balance of mass gives

$$[V_1 s_1 + V_2 s_2]\,\frac{dp}{dt} = \nu s_1 (p_e - p) \quad \text{or} \quad \frac{dp}{dt} = k(p_e - p),$$

where $k = \nu s_1/(V_1 s_1 + V_2 s_2)$ is a constant for the tissue. A simple diagram for this model is presented in **Figure 1**.

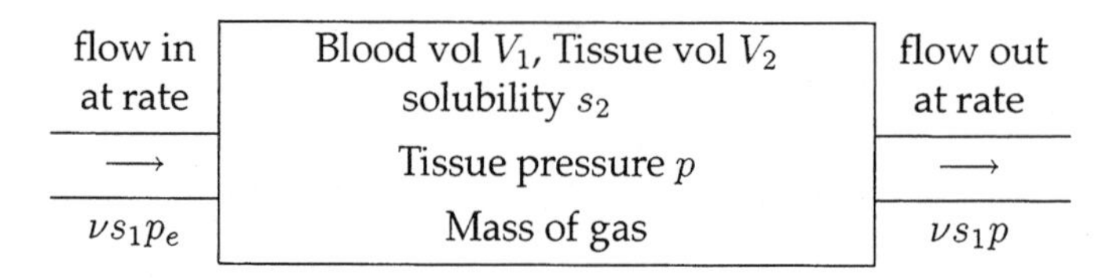

Figure 1. Diagram for the compartment model.

In Haldane's time, this model was thought to be appropriate for both compression ($p_e \geq p$) and decompression ($p \geq p_e$). It was known already that various tissues in the body required different values of s_2, V_1, V_2, and ν, and the same blood does not flow through all tissues. In devising his tables, Haldane considered five different values for the constant k in the differential equation. His calculations were based on solutions of the differential equation and on the experimental result that the external absolute pressure could be reduced by the factor M at any time without an attack of the bends occurring.

In the work that follows, we assume for simplicity that air is all nitrogen. It can be shown that this in fact makes no significant difference to the results (see **Exercise 6**).

4. Solution of the Differential Equation

The differential equation

$$\frac{dp}{dt} = k(p_e - p), \tag{1}$$

where k and p_e are known constants, can be solved to find the pressure p at any time t, provided that the pressure p is known at one instant of time, usually taken to be $t = 0$ (we measure the elapsed time from the instant at which the pressure is known), i.e., $p(0) = p_0$, a known constant. If you know enough integral calculus, you can find the solution of the equation, as shown below, by the method of separation of variables. If you do not know integral calculus, the solution can be verified directly by substitution in **(1)**.

To separate variables, we write **(1)** as

$$\frac{1}{p_e - p}\frac{dp}{dt} = k$$

and integrate (antidifferentiate) both sides with respect to t. This gives

$$\int \frac{1}{p_e - p}\frac{dp}{dt}\,dt = \int \frac{1}{p_e - p}\,dp = \int k\,dt.$$

Performing the integrations, we get

$$-\ln|p_e - p| = kt + c$$

where c is an arbitrary constant. Taking exponentials of both sides gives

$$|p_e - p| = e^{-(kt+c)} = e^{-kt}e^{-c} = Ae^{-kt},$$

where A is an arbitrary constant, $A = e^{-c}$. Since we also require $p(0) = p_0$, it follows that $|p_e - p_0| = A$, and we obtain the solution

$$p = p_e - (p_e - p_0)e^{-kt}. \qquad \textbf{(2)}$$

Graphs of solutions for the case $p_0 = 1$ atm, $p_e = 3$ atm with (a) $k = 0.2\ \text{min}^{-1}$, (b) $k = 0.1\ \text{min}^{-1}$ are given in **Figure 2**. The curves represent the pressure p in the tissues of a diver at time t min after descending from the surface ($p = 1$ atm) to a depth of about 66 ft ($p = 3$ atm). Similarly, graphs for the case $p_0 = 3$ atm, $p_e = 1$ atm with (a) $k = 0.2\ \text{min}^{-1}$, (b) $k = 0.1\ \text{min}^{-1}$ are given in **Figure 3**. Here the curves represent the pressure t minutes after ascending to the surface from a point where the tissue pressure is 3 atm.

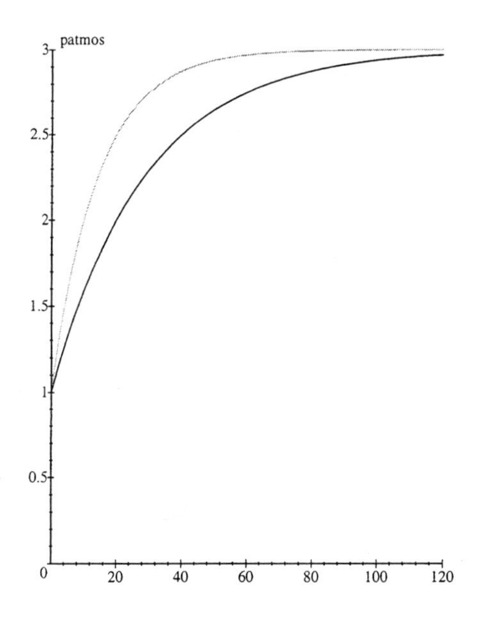

Figure 2. Solutions for the case $p_0 = 1$ atm, $p_e = 3$ atm. The lower curve is for $k = 0.2\ \text{min}^{-1}$ and the upper curve is for $k = 0.1\ \text{min}^{-1}$. The curves give the pressure p in the tissues of a diver at time t min after descending from the surface ($p = 1$ atm) to a depth of about 66 ft ($p = 3$ atm).

The role of the constant k, which is measured in min^{-1} if t is measured in min, is indicated in **Figures 2** and **3**. When p_0 and p_e are held constant, it takes

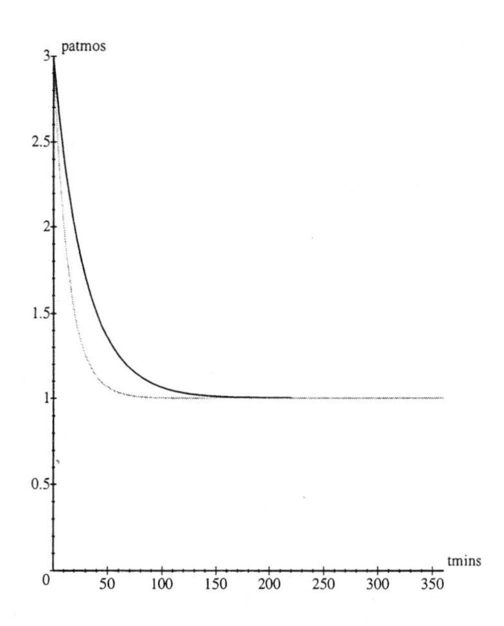

Figure 3. Solutions for the case $p_0 = 3$ atm, $p_e = 1$ atm. The lower curve is for $k = 0.2$ min^{-1} and the upper curve is for $k = 0.1$ min^{-1}. The curves give the pressure t min after ascending to the surface from a point where the tissue pressure is 3 atm.

twice as long to attain a given pressure when $k = 0.1$ as it does when $k = 0.2$. We also see that for any positive k, p approaches the constant external pressure p_e as t becomes large ($t \to \infty$) no matter what the value of p_0. In other words, the pressure equalizes over time, as expected.

5. The Half-Time

Because solutions of the exponential nature of **(2)** all have the same asymptote $p = p_e$ for all positive values of k, they are often characterized by their half-time, or half-life as it is called in the case of radioactive decay.

The half-time is the time required for the difference between p and the external pressure p_e to drop to exactly one half of its original value, that is, the time at which $(p - p_e) = (p_0 - p_e)/2$.

From **(2)**, we see that if T is the half-time, then

$$p - p_e = (p_0 - p_e)e^{-kT} = \frac{1}{2}\,(p_0 - p_e)$$

and hence

$$e^{-kT} = \frac{1}{2} \Rightarrow e^{kT} = 2 \Rightarrow kT = \ln 2. \qquad \textbf{(3)}$$

From this equation we see that $k = \ln 2/T$ no matter what the values of p_0 and p_e are, and that the half-time T for a tissue completely determines the value of k in **(2)**. This makes the half-time extremely useful in characterizing the various tissues in the body.

The relationships between bottom times and decompression programmes differ for different half-times. The human body contains many different tissues, as Haldane knew, and a safe decompression programme must make sure that the bends do not occur in any of them. Haldane did not have exact values for half-times, so to compile his tables he used five different values (5, 10, 20, 40, and 75 min) in the belief that this would cover any reasonable spectrum of half-times. His tables were successful over the wide range of dives undertaken at that time and for some considerable time thereafter.

Noting that

$$e^{-kT} = \frac{1}{2}, \quad e^{-kt} = e^{-kT\left(\frac{t}{T}\right)} = \left(e^{-kT}\right)^{(t/T)} = \left(\frac{1}{2}\right)^{t/T},$$

we rewrite **(2)** as

$$p = p_e + (p_0 - p_e)\left(\frac{1}{2}\right)^{t/T}. \qquad \textbf{(4)}$$

6. Scuba and No-Stop Dives

Most recreational divers usually dive to a given depth, remain at (or above) that depth for a certain time, and then ascend directly to the surface. This is the "no stop" or "no decompression" dive, as shown in **Table 1** below. The time allowed at the bottom depends on the depth of the dive. For example, the table says that you may stay ("stay" includes descent and ascent) at 70 ft for 50 min.

Table 1.
Diving table (from Hammes and Zimos [1988]).

Depth (ft)	40	50	60	70	80	90	100	110	120	130
Time (min)	200	100	60	50	40	30	25	20	15	10

A no-stop diving table can be produced from our model in the following manner.

We wish to model a situation in which a diver starts with an initial gas tissue pressure of 1 atm and wishes to stay at a depth d ft where the external pressure is $p_e = 1 + d/33$ (33 ft of water gives a pressure of 1 atm; the equation contains a 1 because there is already a pressure of 1 atm at the surface $d = 0$). We use **(2)** to tell us the tissue gas pressure after t min, which will be

$$p = 1 + \frac{d}{33} - \frac{d}{33}e^{-kt} \qquad (k \text{ being known for the given tissue}).$$

Haldane's decompression experiment says that the diver may ascend directly to the surface where the pressure is 1 atm provided that the pressure p attained in the tissues is less than 2.15 atm. Thus the diver has a limiting dive time t_d given by

$$2.15 = 1 + \frac{d}{33}\left(1 - e^{-kt_d}\right),$$
$$\frac{d}{33} = \frac{1.15}{1 - e^{-kt_d}}.$$

This relation gives the time for the tissue as characterized by its value of k (equivalently, by its half-time $T = \ln 2/k$).

The allowable time t_d becomes longer as k becomes less, that is, as the half-time $T(= \ln 2/k)$ becomes greater. To be safe for all tissues, t_d is limited by the tissue with the shortest half-time, which is 5 min in Haldane's scheme. This would give the relation

$$d = \frac{38}{1 - \exp(-t_d \ln 2/5)}.$$

Tables are usually written with t_d as a function of depth d, which our model gives as

$$t_d = \frac{5\ln\left(\frac{d}{d-38}\right)}{\ln 2}.$$

You will find that this relation gives qualitative agreement with published tables (see **Figure 4**); but the quantitative agreement is not very good, because of the conservative nature of Haldane's value of M and his tissue half-time of 5 min for short dives.

7. Dives with Decompression Stops

For dives that fall outside the no-stop dive range, a more complicated set of conditions must be satisfied. Again we follow Haldane's recipe.

The standard method to calculate a decompression routine is to consider a series of stops at depths that are multiples of 10 ft. The first stop must be such that the external pressure at that depth is not less than M times the pressure in each of the tissues that has been reached during the stay at the diving depth. The tissue pressures depend on the time spent at depth and on the tissue half-time. The greatest tissue pressure will be in the tissue with the shortest half-time. Consider the following three examples.

Example 1. Consider a one-hour dive at a depth of 66 ft, where the pressure is approximately 3 atm. To save some calculations, we assume

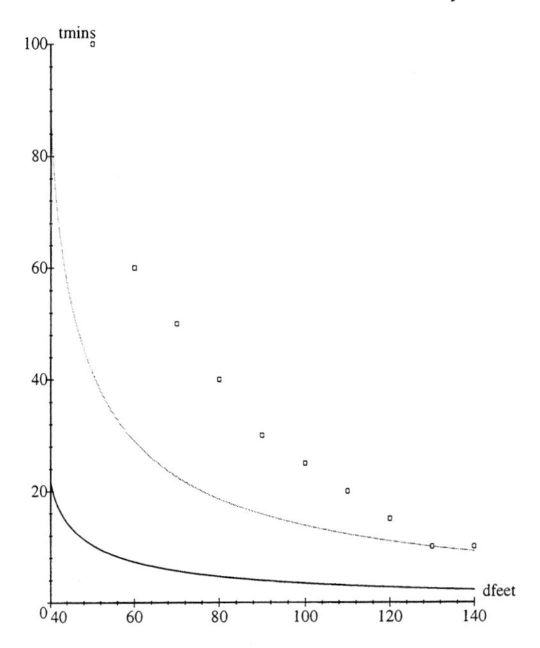

Figure 4. No-stop dive. Graph of $t_d = 5\ln\left(\frac{d}{d-38}\right)/\ln 2$ and $t_d = 20\ln\left(\frac{d}{d-38}\right)/\ln 2$, compared with points from the diving table of **Table 1**.

that there are three tissues (as opposed to Haldane's five) with half-times 10, 20, and 40 min. From **(4)**, the pressure of a tissue at an external pressure p_e is

$$p = p_e + (p_0 - p_e)\left(\frac{1}{2}\right)^{t/T},$$

where p_0 is the initial tissue pressure, T the tissue half-time, and t is the length of time at depth (in minutes). The pressure p_0 at the beginning of the dive is 1 atm. After one hour at 66 ft (3 atm for 60 min, $p_e = 3$ atm), tissue pressures are

$$\begin{aligned}
T &= 10\text{-min tissue}: & p &= 3 - 2\left(\tfrac{1}{2}\right)^{6} & &\approx 2.97\\
T &= 20\text{-min tissue}: & p &= 3 - 2\left(\tfrac{1}{2}\right)^{3} & &\approx 2.75\\
T &= 40\text{-min tissue}: & p &= 3 - 2\left(\tfrac{1}{2}\right)^{3/2} & &\approx 2.29.
\end{aligned}$$

It is safe to ascend to an external pressure of $2.97/2.15 = 1.35$, or about 12.5 ft. To keep the ascent steps in multiples of 10 ft, the first ascent is made to 20 ft (1.60 atm).

At this point, the diver makes a stop. We have to decide how long this stop should be. To do this, we must decide the depth for the next stop. We choose 10 ft or 1.30 atm. The diver must remain at 20 ft until all tissue pressures have declined to a value that will be safe when the diver ascends to 1.30 atm—that is, until all tissue pressures are reduced to $2.15 \times 1.30 = 2.795$ atm. The three tissue pressures at the beginning of the 20-ft stop are 2.97, 2.75, 2.29. The pressures of the 20-min and 40-min tissues are already low enough to ascend to 10 ft. For the 10-min tissue, t min will result in pressures

$$\text{10-min tissue:} \qquad p = 1.6 + 1.37 \left(\frac{1}{2}\right)^{t/10}.$$

(Again we are using **(4)**, with $p_e = 1.6$ and $p_0 = 2.97$ for $T = 10$.) The diver must remain at the 20-ft level until all tissue pressures are below the pressure 2.795 atm that is safe at the 10-ft stop (1.3 atm). For the 10-min tissue, this means t must be greater than the solution of

$$2.975 = 1.6 + 1.37 \left(\frac{1}{2}\right)^{t/10} \quad \text{or} \quad t = 10 \ln \left(\frac{1.37}{1.195}\right) / \ln 2 \approx 1.971.$$

Suppose that we make a 2-min stop at 20 ft. We must calculate the tissue pressures after 2 min at 20 ft:

$$\begin{aligned} \text{10-min:} \quad & p = 1.6 + 1.37 \left(\tfrac{1}{2}\right)^{.2} && = 2.79 \\ \text{20-min:} \quad & p = 1.6 + 1.15 \left(\tfrac{1}{2}\right)^{.1} && = 2.67 \\ \text{40-min:} \quad & p = 1.6 + .69 \left(\tfrac{1}{2}\right)^{.05} && = 2.27. \end{aligned}$$

These are the initial pressures at the 10-ft (1.3-atm) stop. The next ascent will be to the surface (1 atm), where the safe pressure will be 2.15. The stop at 10 ft (1.3 atm) must be long enough that all three pressures will drop below 2.15. For a stop of t min, the pressures will be

$$\begin{aligned} \text{10-min:} \quad & p = 1.3 + 1.49 \left(\tfrac{1}{2}\right)^{t/10} \\ \text{20-min:} \quad & p = 1.3 + 1.37 \left(\tfrac{1}{2}\right)^{t/20} \\ \text{40-min:} \quad & p = 1.3 + .97 \left(\tfrac{1}{2}\right)^{t/40}, \end{aligned}$$

and t must be large enough that all three are less than 2.15. For the 10-min tissue, this requires 7.62 min, for the 20-min tissue 13.77 min, and for the 40-min tissue 8.09 min. The stop at 10 ft must be greater than 13.77 min—say 14 min. An appropriate decompression procedure for a one-hour dive at 66 ft would feature stops of

2 min	at 20 ft
14 min	at 10 ft.

The ascent would also be lengthened by the time to ascend the 66 ft, about 1.5 min.

Example 2. We take an ascent as recommended in Haldane's tables [Hempleman 1982, 330]. For a dive of 130 min at 90 ft, Haldane's tables recommend stops of

5 min	at	30 ft
25 min	at	20 ft
30 min	at	10 ft.

In this calculation, we will use all five of Haldane's half-times of 5, 10, 20, 40, and 75 min.

First, we calculate the saturation levels for a dive of 130 min at 90 ft ≈ 3.73 atm. Then we calculate the pressures at the end of the period spent at each stopping point. Finally, we note the safe pressure to ascend to the next stop (see **Table 2**).

Table 2.

Analysis of ascent recommended by Haldane for a 130-min dive at 90 ft.

Tissue	Pressure			
half-time (min)	90 ft = 3.73 atm	30 ft = 1.9 atm	20 ft = 1.6 atm	10 ft = 1.3 atm
5	3.73	2.82	1.64	1.30
10	3.73	3.19	1.88	1.37
20	3.70	3.41	2.36	1.67
40	3.44	3.31	2.71	2.14
75	2.91	2.86	2.60	2.285
Safe pressure at next stop	4.08	3.44	2.8	2.15

We see that at every stage except one, a safe pressure is attained in each tissue to allow the diver to ascend to the next stop. The exception is the last ascent to the surface for the 75-min tissue. Haldane allowed 2 min to move to and from the stops; if this time were included, the final pressures would be slightly reduced. This example, however, shows a problem with Haldane's tables for long dives.

A much more recent U.S. Navy Table T–10 (reproduced in Hammes and Zimos [1988]) gives for this dive stopping times of

5 min	at	30 ft
36 min	at	20 ft
74 min	at	10 ft.

This decompression procedure allows for even larger half-times than 75 min.

Figure 5 shows graphs of the tissue pressures for half-times of 5, 10, 20, 40, and 75 min, using the decompression scheme from Haldane's tables. The piecewise "step" graph at the right indicates the safe pressure at the stops.

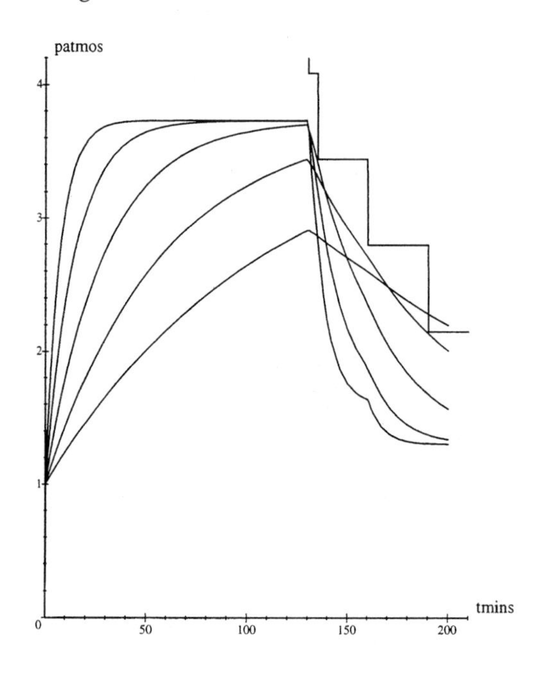

Figure 5. A 130-min dive to 90 ft followed by ascent with decompression stops as recommended by Haldane's tables. At left, from top to bottom, are tissue pressures at 90 ft for half-times of 5, 10, 20, 40, and 75 min. At right, from top to bottom, are the tissue pressures during ascent. The piecewise "step" graph at far right indicates the safe pressures at the stops.

Example 3. We consider a dive to 80 ft = 3.43 atm for one hour. Haldane's tables give stops of

9 min	at	20 ft
18 min	at	10 ft.

Again we give the pressures as the diver *leaves* each level to proceed to the next (see **Table 3**).

In this case, a safe tissue pressure has been reached at all levels for all tissues before proceeding. This decompression procedure, however, is now considered to be rather conservative. The U.S. Navy table suggests 17 min at 10 ft as the only stop for this dive.

A procedure of this kind can be calculated with as many tissues as appropriate. (You might like to write a computer programme to carry out the steps.)

Table 3.
Analysis of ascent recommended by Haldane for a 60-min dive at 80 ft.

Tissue	Pressure 80 ft = 3.43 atm	20 ft = 1.6 atm	10 ft = 1.3 atm
5	3.43	2.13	1.37
10	3.39	2.56	1.66
20	3.17	2.75	2.08
40	2.57	2.43	2.13
75	2.03	2.00	1.89
Safe pressure at next stop	3.44	2.80	2.15

Exercises

In all exercises, assume that $M = 1/2.15$.

1. Find a decompression procedure for a dive of 40 min at 3.5 atm (80–85 ft) with stops at 1.7 atm (23 ft) and 1.3 atm (10 ft). (Consider only 10- and 20-min tissues.)

2. Find a decompression procedure for a 2-hr dive at 4.0 atm (100 ft) with stops at 1.9 atm (30 ft), 1.6 atm (20 ft), and 1.3 atm (10 ft). (Consider 10-, 20-, and 40-min tissues.)

3. Show that a slightly faster ascent for the dive of **Exercise 2** could be made if three stops of equal duration T_1 are made, the first at 1.9 atm (30 ft) and the second and third at depths to be determined. (As a first step, consider only the 40-min tissue; then verify that the steps are appropriate for the 10-min and 20-min tissues.)

4. Show that for a single tissue half-time T and an n-stop decompression schedule, the shortest total ascent time is achieved by using equal times at each step and determining the depths of each step according to the time. (The actual time at each step is determined by the number of steps.)

5. Show that for a single tissue, it is possible to have a continuous ascent in which the tissue pressure at time t is exactly 2.15 times the external pressure that the diver is experiencing at that time. Find the diver's depth at time t (pressure $= 1 + d/33$ atm, where d is in feet). Using such a scheme, find how long it would take to ascend from a long dive at 4 atm. (Assume a single tissue of half-time 40 min and an instantaneous ascent from 4 atm to $1.86 = 4/2.15$ atm.)

6. If the nitrogen (partial) pressure in a tissue is 80% of the pressure, and the safe nitrogen pressure for a no-stop dive is 2.15 times that of the nitrogen partial pressure in the atmosphere (0.8 atm), show that the equation relating time and depth for no-stop dives is unaltered.

7. Check for safety the following recommendations from Haldane's tables for a dive of 45 min at 85 ft (3.58 atm). Stop 2 min at 30 ft, 7 min at 20 ft, 15 min at 10 ft. (U.S. Navy Table T–10 [Hammes and Zimos 1988] gives one stop of 17 min at 10 ft for this dive.)

8. Repetitive Dives

A major portion of the scuba diving tables is devoted to repetitive diving. The problem with repetitive diving is the fact that after one "no decompression" dive, the tissue pressure may be 2.15 times atmospheric pressure. An immediate dive back to a depth greater than 37 ft (external pressure greater than 2.15 atm) would raise the tissue pressure to above the limit that would allow a safe ascent to the surface. A break at the surface between dives lessens the pressure when the second dive is commenced, but it takes about twelve hours to restore all tissue pressures to 1 atm. The tissue pressure remaining after the first dive is known as the *residual nitrogen pressure* (RNP). We consider only a 20-min tissue in making our calculations, to keep things simple.

Example 4. Dive (1): 15 min at 80 ft. Dive (2) is to be to a depth of 100 ft after a one-hour break at the surface. We calculate the safe time for a "no decompression" second dive (20-min tissue only).

Tissue pressure p after 15 min at 80 ft ($p_e \approx 3.4$ atm):

$$p = 3.4 - 2.4\left(\frac{1}{2}\right)^{3/4} = 1.97.$$

Since this is less than 2.15, it is safe to ascend to the surface.

Tissue pressure p after one hour at the surface ($p_e = 1$ atm):

$$p = 1 + .97\left(\frac{1}{2}\right)^{3} = 1.12.$$

Descent to 100 ft (4 atm):

$$p = 4 - 2.88\left(\frac{1}{2}\right)^{t/20}.$$

The diver may remain until $p = 2.15$, that is, until

$$t = 20\ln(2.88/1.85)/\ln 2 = 12.77 \text{ min}.$$

Figure 6 shows the pressure as a function of time for this example.

Actual scuba tables cover the large numbers of different calculations by classifying the residual nitrogen pressures into groups A, B, C, etc. The group

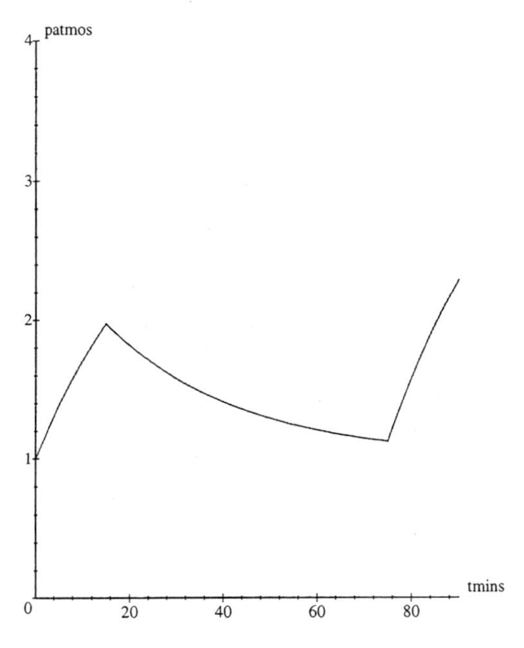

Figure 6. Repetitive dive.

is found after the first dive. The effect of remaining at the surface for a given time period is to change the group; the new group determines the safe time for the next dive. We give an example.

Example 5.

Dive 1: 100 ft for 15 min
Dive 2: 80 ft
Time at surface between dives: 1 hr

We consult **Table 4**. First look at the row for a dive to 100 ft. Note that the no-stop time is 25 min. Our dive is for 15 min, so we go across the row until we reach 15. We then move down the corresponding column and find the repetitive group label "E".

The stay at the surface is for 60 min. We continue along the column until we come to the two numbers that bracket 60 min:

$$0:55$$
$$1:57.$$

We now proceed left acrooss this row until we find a new repetitive group label "D".

For the second dive, at depth 80 ft, we use the label D. We continue across the row until we reach the column corresponding to 80 ft (at right).

The entry contains the numbers 18 (RNT) and 22 (TR). This means that because of the previous dive, it is as if we had already been at this depth for 18 min, and our time remaining is 22 min. We must be back at the surface within 22 min.

Exercises

Use **(2)** in the following exercises.

8. Consider the same sequence of dives as in **Example 4** but include a 40-min tissue. Does this make a difference for the second stop time?

9. Find the safe time for a second dive to 80 ft one hour after a first dive to 100 ft for 10 min. Consider tissue half-times of 20 min and 40 min.

9. Changes in Pressure During Descent and Ascent

To this point, we have assumed that the passage from one level to another is instantaneous. This is not possible; moreover, rapid motion is not recommended. A steady ascent or descent rate of about 60 ft/min is not unreasonable, and we will now examine the effect on tissue pressure of ascending at such a rate.

Our basic equation

$$\frac{dp}{dt} = k(p_e - p)$$

(where p_e is the external pressure) still holds, but p_e is no longer constant. For a descent at a constant rate of 60 ft/min, we have $p_e = 1 + 60t/33$ atm, and the differential equation becomes

$$\frac{dp}{dt} = k\left(1 + \frac{60t}{33} - p\right) \qquad \text{or} \qquad \frac{dp}{dt} + kp = k\left(1 + \frac{60t}{33}\right). \tag{5}$$

This is no longer a separable equation but a first-order linear equation, and it must be solved in a different manner. Here we describe one possible method.

First we try to guess a solution. After examining the equation, we feel that $p = A + Bt$, where A and B must be selected, seems a possible guess. If we substitute this into **(5)**, we see that we get a solution if we can choose A, B so that

$$B + k(A + Bt) = k\left(1 + \frac{60t}{33}\right).$$

Table 4.

Diving table. ©NAUI 1987. Reproduced by permission.

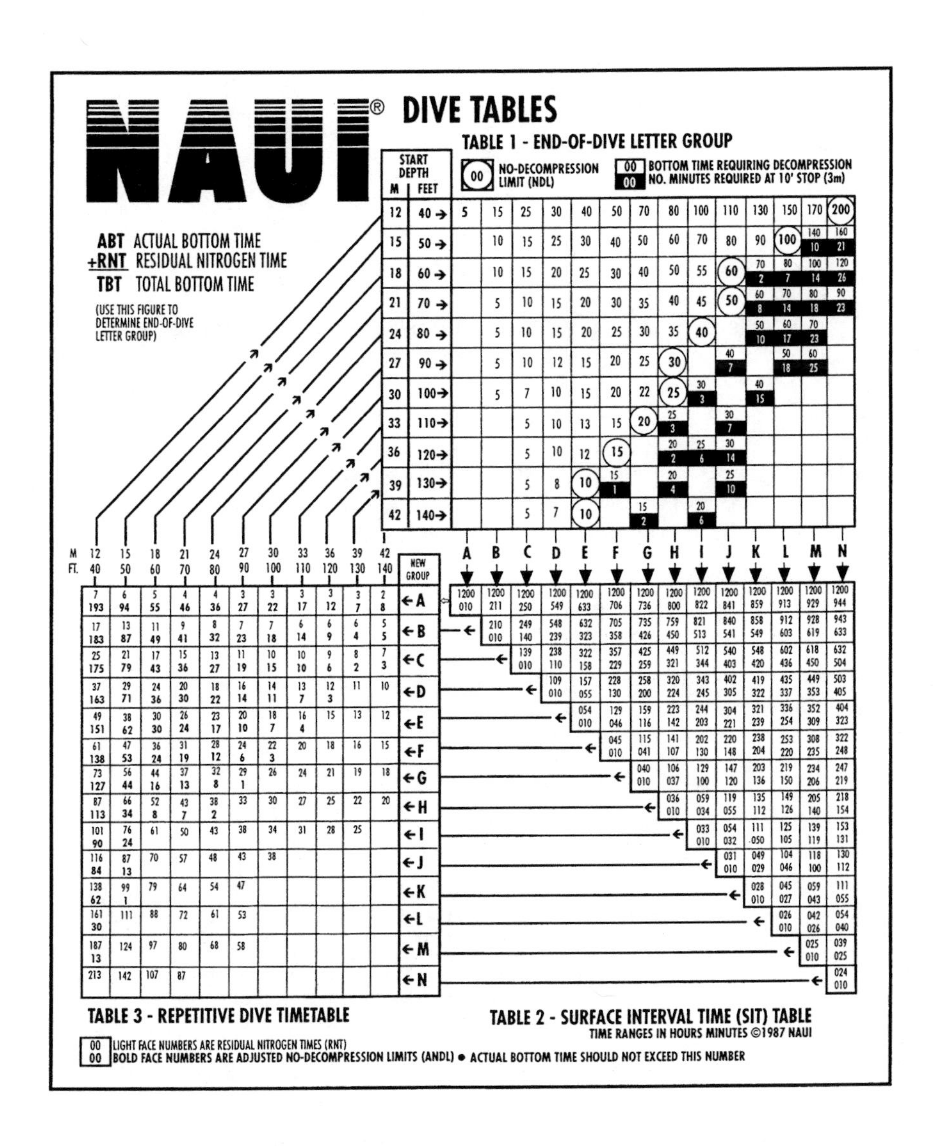

NAUI® DIVE TABLES

ABT ACTUAL BOTTOM TIME
+RNT RESIDUAL NITROGEN TIME
TBT TOTAL BOTTOM TIME

(USE THIS FIGURE TO DETERMINE END-OF-DIVE LETTER GROUP)

TABLE 1 - END-OF-DIVE LETTER GROUP

(00) NO-DECOMPRESSION LIMIT (NDL) — 00 / 00 BOTTOM TIME REQUIRING DECOMPRESSION / NO. MINUTES REQUIRED AT 10' STOP (3m)

START DEPTH M	FEET	A	B	C	D	E	F	G	H	I	J	K	L	M	N
12	40 →	5	15	25	30	40	50	70	80	100	110	130	150	170	(200)
15	50 →		10	15	25	30	40	50	60	70	80	90	(100)	140 / 10	160 / 21
18	60 →		10	15	20	25	30	40	50	55	(60)	70 / 2	80 / 7	100 / 14	120 / 26
21	70 →		5	10	15	20	30	35	40	45	(50)	60 / 8	70 / 14	80 / 18	90 / 23
24	80 →		5	10	15	20	25	30	35	(40)		50 / 10	60 / 17	70 / 23	
27	90 →		5	10	12	15	20	25	(30)		40 / 7		50 / 18	60 / 25	
30	100 →		5	7	10	15	20	22	(25)	30 / 3		40 / 15			
33	110 →			5	10	13	15	(20)	25 / 3		30 / 7				
36	120 →			5	10	12	(15)		20 / 2	25 / 6	30 / 14				
39	130 →			5	8	(10)	15 / 1		20 / 4		25 / 10				
42	140 →			5	7	(10)		15 / 2		20 / 6					

TABLE 2 - SURFACE INTERVAL TIME (SIT) TABLE

TIME RANGES IN HOURS MINUTES ©1987 NAUI

NEW GROUP	A	B	C	D	E	F	G	H	I	J	K	L	M	N
← A	1200 / 010	1200 / 211	1200 / 250	1200 / 549	1200 / 633	1200 / 706	1200 / 736	1200 / 800	1200 / 822	1200 / 841	1200 / 859	1200 / 913	1200 / 929	1200 / 944
← B		210 / 010	249 / 140	548 / 239	632 / 323	705 / 358	735 / 426	759 / 450	821 / 513	840 / 541	858 / 549	912 / 603	928 / 619	943 / 633
← C			139 / 010	238 / 110	322 / 158	357 / 229	425 / 259	449 / 321	512 / 344	540 / 403	548 / 420	602 / 436	618 / 450	632 / 504
← D				109 / 010	157 / 055	228 / 130	258 / 200	320 / 224	343 / 245	402 / 305	419 / 322	435 / 337	449 / 353	503 / 405
← E					054 / 010	129 / 046	159 / 116	223 / 142	244 / 203	304 / 221	321 / 239	336 / 254	352 / 309	404 / 323
← F						045 / 010	115 / 041	141 / 107	202 / 130	220 / 148	238 / 204	253 / 220	308 / 235	322 / 248
← G							040 / 010	106 / 037	129 / 100	147 / 120	203 / 136	219 / 150	234 / 206	247 / 219
← H								036 / 010	059 / 034	119 / 055	135 / 112	149 / 126	205 / 140	218 / 154
← I									033 / 010	054 / 032	111 / .050	125 / 105	139 / 119	153 / 131
← J										031 / 010	049 / 029	104 / 046	118 / 100	130 / 112
← K											028 / 010	045 / 027	059 / 043	111 / 055
← L												026 / 010	042 / 026	054 / 040
← M													025 / 010	039 / 025
← N														024 / 010

TABLE 3 - REPETITIVE DIVE TIMETABLE

NEW GROUP	12 M / 40 FT.	15 / 50	18 / 60	21 / 70	24 / 80	27 / 90	30 / 100	33 / 110	36 / 120	39 / 130	42 / 140
← A	7 / **193**	6 / **94**	5 / **55**	4 / **46**	4 / **36**	3 / **27**	3 / **22**	3 / **17**	3 / **12**	3 / **7**	2 / **8**
← B	17 / **183**	13 / **87**	11 / **49**	9 / **41**	8 / **32**	7 / **23**	7 / **18**	6 / **14**	6 / **9**	6 / **4**	5 / **5**
← C	25 / **175**	21 / **79**	17 / **43**	15 / **36**	13 / **27**	11 / **19**	10 / **15**	10 / **10**	9 / **6**	8 / **2**	7 / **3**
← D	37 / **163**	29 / **71**	24 / **36**	20 / **30**	18 / **22**	16 / **14**	14 / **11**	13 / **7**	12 / **3**	11	10
← E	49 / **151**	38 / **62**	30 / **30**	26 / **24**	23 / **17**	20 / **10**	18 / **7**	16 / **4**	15	13	12
← F	61 / **138**	47 / **53**	36 / **24**	31 / **19**	28 / **12**	24 / **6**	22 / **3**	20	18	16	15
← G	73 / **127**	56 / **44**	44 / **16**	37 / **13**	32 / **8**	29 / **1**	26	24	21	19	18
← H	87 / **113**	66 / **34**	52 / **8**	43 / **7**	38 / **2**	33	30	27	25	22	20
← I	101 / **90**	76 / **24**	61	50	43	38	34	31	28	25	
← J	116 / **84**	87 / **13**	70	57	48	43	38				
← K	138 / **62**	99 / **1**	79	64	54	47					
← L	161 / **30**	111	88	72	61	53					
← M	187 / **13**	124	97	80	68	58					
← N	213	142	107	87							

00 LIGHT FACE NUMBERS ARE RESIDUAL NITROGEN TIMES (RNT)
00 BOLD FACE NUMBERS ARE ADJUSTED NO-DECOMPRESSION LIMITS (ANDL) • ACTUAL BOTTOM TIME SHOULD NOT EXCEED THIS NUMBER

The choice $B = 60/33$, with $kA + B = k$, hence $A = 1 - 60/33k$ gives a solution

$$p = 1 + \frac{60}{33}\left(t - \frac{1}{k}\right).$$

We call this a *particular* integral. If we then write

$$u = p - \left[1 + \frac{60}{33}\left(t - \frac{1}{k}\right)\right] = p - 1 - \frac{60}{33}t + \frac{60}{33k},$$

where p is any solution of the equation, it follows that

$$\frac{du}{dt} + ku = \frac{dp}{dt} + kp - \frac{60}{33} - k\left(1 + \frac{60t}{33}\right) + \frac{60}{33} = 0,$$

since p is a solution of **(5)**.

If $du/dt + ku = 0$, then we can again use separation of variables to get

$$\int \frac{1}{u}\frac{du}{dt}\,dt = \int k\,dt$$

which implies that $-\ln|u| = kt + C$, or $u = Ae^{-kt}$, where A is an arbitrary constant. In this approach, u is usually called the *complementary function*. Thus, if p is any solution of **(5)**, it can be written as

$$p = 1 + \frac{60}{33}\left(t - \frac{1}{k}\right) + u = 1 + \frac{60}{33}\left(t - \frac{1}{k}\right) + Ae^{-kt};$$

that is, any solution is the sum of a particular integral and a complementary function. The technique may be used on any first-order linear equation. To satisfy an initial condition $p(0) = p_0$, we get

$$1 - \frac{60}{33k} + A = p_0 \quad \text{or} \quad A = p_0 - 1 + \frac{60}{33k},$$

$$p = 1 + \frac{60t}{33} - \frac{60}{33k} + \left(p_0 - 1 + \frac{60}{33k}\right)e^{-kt} \qquad \textbf{(6)}$$

$$= 1 + \frac{60t}{33} - \frac{60}{33k} + \left(p_0 - 1 + \frac{60}{33k}\right)\left(\frac{1}{2}\right)^{(t/T)}. \qquad \textbf{(7)}$$

A similar solution could be obtained for an ascent from a given depth.

Example 6. Find the pressure in a 20-min tissue on arrival at a depth of 100 ft (4 atm) after a descent from the surface at a rate of 60 ft/min.

The time to descend 100 ft at 60 ft/min is $10/6 = 5/3$ min.

The initial pressure is $p_0 = 1$, and $k = \ln 2/T = .03466$.

Therefore,

$$p = 1 + \frac{60}{33} \cdot \frac{5}{3} - \frac{60}{33(.03466)} + \frac{60}{33(0.3466)} \left(\frac{1}{2}\right)^{1/12} = 1.086.$$

To do a complete dive, we would have to include these changes of pressure in the complete diving schedule. We will not do this, although it is merely tedious rather than difficult.

We note finally that if the descent had been considered instantaneously, the pressure after $5/3$ min at a depth of 100 ft would be 1.17 atm.

Exercise

10. Find the tissue pressure for a 20-min tissue at the end of an ascent from 100 ft to 10 ft at a speed of 60 ft/min, assuming that the pressure at the beginning of the ascent was 4 atm. Compare it with the pressure at 10 ft after an instantaneous ascent.

10. Conclusion

In this Module, we have discussed a simple technique for derivation of diving tables, which is based on a model proposed by Haldane. Although modern diving tables cannot be devised by means of such simple techniques, most of them have been developed by refinements to the simple model and methods proposed by Haldane, as tempered by experience (see, for example, Bornmann [1970]).

11. Solutions to the Exercises

All solutions use either **(4)** or its inverse:

$$p = p_e + (p_0 - p_e)\left(\frac{1}{2}\right)^{t/T} \qquad \text{or} \qquad t = T \ln\left(\frac{p_0 - p_e}{p - p_e}\right) / \ln 2.$$

1. During the dive, $p_e = 3.5$, $t = 40$ min, $p_0 = 1$.
For $T = 10$, $p = 3.344$; for $T = 20$, $p = 2.875$. It is safe to ascend to 20 ft = 1.6 atm, because $2.15 \times 1.6 = 3.44$. The stop at 1.6 should be long enough that an ascent to 1.3 will be safe. This requires that p be reduced to $2.15 \times 1.3 = 2.795$.

For $T = 10$, this requires $t = 10 \ln\left(\frac{3.344 - 1.6}{2.795 - 1.6}\right) / \ln 2 \simeq 5.454$ min ($p_0 = 3.344$, $p_e = 1.6$).

For $T = 20$, this requires $t = 20 \ln\left(\dfrac{2.875 - 1.6}{2.795 - 1.6}\right) / \ln 2 \simeq 1.870$ min ($p_0 = 2.875$, $p_e = 1.6$).

Thus, a stop of 5.454 min is required. After 5.454 min, the pressure in the $T = 20$ tissue is 2.655, and that in the $T = 10$ tissue is 2.795.

The stop at 1.3 (10 ft) should be long enough that an ascent to the surface (1 atm) is safe. This requires that p be reduced to 2.15.

For $T = 10$, this requires $t = 10 \ln\left(\dfrac{2.795 - 1.3}{2.15 - 1.3}\right) / \ln 2 \simeq 8.146$ min.

For $T = 20$, this requires $t = 20 \ln\left(\dfrac{2.655 - 1.3}{2.15 - 1.3}\right) / \ln 2 \simeq 13.455$ min.

A safe schedule is then a 5.454-min stop at 1.6 (20 ft) and a 13.455-min stop at 10 ft. The total stopping time is 18.909 min.

2. By similar means as in **Exercise 1**, the pressures at the end of the dive where $p_e = 4$, $p_0 = 1$, $t = 120$ are: for $T = 10$, $p = 4$; for $T = 20$, $p = 3.953$; for $T = 40$, $p = 3.625$.

   Stop 1 at 1.9 *atm*: (This is safe since since $1.9 \times 2.15 = 4.085$.) Times to reduce pressure to $1.6 \times 2.15 = 3.44$ are: for $T = 10$, 4.47 min; for $T = 20$, 8.296 min; for $T = 40$, 6.547 min.

   A stop of 8.296 min is required. After this stop, the $T = 10$ tissue will have a pressure below that of the $T = 20$ tissue, and this will remain true for the rest of the dive. We need not consider the $T = 10$ tissue further.

   After 8.296 min at 1.9, $T = 20$ has pressure 3.44 and $T = 40$ has pressure 3.39.

   Stop 2 at 1.6 *atm*: Times to reduce pressure to $2.15 \times 1.3 = 2.795$ are: for $T = 20$, 12.5 min; for $T = 40$, 23.4 min. From this point on we need only consider the $T = 40$ tissue. After stop 2, its pressure is 2.795.

   Stop 3 at 1.3 *atm*: Time to reduce pressure to 2.15 is 32.584 min for the $T = 40$ tissue.

   The total time for all stops is 64.3 min.

3. The same dive as in **Exercise 2**. We consider the 40-min tissue only and make three stops of equal time. The first stop is at 1.9, but the depth of the remaining stops must be calculated from the condition of equal times.

   After the dive, the pressure in the $T = 40$ tissue is 3.625. Ascent to 1.9 is certainly safe.

   Suppose that the second and third stops are at pressures p_2, p_3. Then the diver must stay at 1.9 until $p = 2.15p_2$, must stay at p_2 until $p = 2.15p_3$, and must stay at p_3 until $p = 2.15$. From the inverse of **(4)**, the equalization times are

$$t_1 = \frac{40}{\ln 2} \ln\left(\frac{3.625 - 1.9}{2.15p_2 - 1.9}\right) = \frac{40}{\ln 2} \ln\left(\frac{2.15p_2 - p_2}{2.15p_3 - p_2}\right)$$
$$= \frac{40}{\ln 2} \ln\left(\frac{2.15p_3 - p_3}{2.15 - p_3}\right).$$

This gives

$$\frac{1.725}{2.15p_2 - 1.9} = \frac{1.15}{2.15\dfrac{p_3}{p_2} - 1} = \frac{1.15}{2.15\left(\dfrac{1}{p_3}\right) - 1}.$$

The last two equations give $p_3/p_2 = 1/p_3$, or $p_2 = p_3^2$. The first two then give $2.4725p_3^3 - 0.46p_3 - 3.70875 = 0$. The only real positive solution is $p_3 = 1.199$. Thus, $p_2 = p_3^2 = 1.438$ and $t_1 = 21.438$. The total stop time is $3t_1 = 64.314$, a very small improvement. We can verify that after the first stop, the $T = 20$ pressure is 2.877 and the $T = 10$ pressure is 2.375, both below the 3.092 ($= 2.15 \times 1.438$) of the $T = 40$.

4. We assume that the tissue pressure at the beginning of the ascent is p_0, which is known. The three stops will be at pressures p_1, p_2, p_3, where $p_1 = p_0/2.15$ and the pressures at the ends of the stops will be $2.15p_2$, $2.15p_3$, 2.15. The times at each stop will then be

$$t_1 = \frac{T}{\ln 2} \ln\left(\frac{p_0 - p_1}{2.15p_2 - p_1}\right) = \frac{T}{\ln 2} \ln\left(\frac{1.15}{2.15\,\dfrac{p_2}{p_1} - 1}\right),$$

$$t_2 = \frac{T}{\ln 2} \ln\left(\frac{2.15p_2 - p_2}{2.15p_3 - p_2}\right) = \frac{T}{\ln 2} \ln\left(\frac{1.15}{2.15\,\dfrac{p_3}{p_2} - 1}\right),$$

$$t_3 = \frac{T}{\ln 2} \ln\left(\frac{2.15p_3 - p_3}{2.15 - p_3}\right) = \frac{T}{\ln 2} \ln\left(\frac{1.15}{2.15\left(\dfrac{1}{p_3}\right) - 1}\right).$$

We wish to minimize $t_1 + t_2 + t_3$ by choosing p_2 and p_3. This is equivalent to maximizing

$$F(p_2, p_3) = \ln\left(\frac{p_2}{p_1} - M\right) + \ln\left(\frac{p_3}{p_2} - M\right) + \ln\left(\frac{1}{p_3} - M\right),$$

where $M = 1/2.15$ and p_1 is known. Using calculus, we find

$$\frac{\partial F}{\partial p_2} = \frac{1}{\left(\dfrac{p_2}{p_1} - M\right)}\frac{1}{p_1} + \frac{1}{\left(\dfrac{p_3}{p_2} - M\right)}\left(\frac{-p_3}{p_2^2}\right) = 0,$$

$$\frac{\partial F}{\partial p_3} = \frac{1}{\left(\dfrac{p_3}{p_2} - M\right)}\frac{1}{p_2} + \frac{1}{\left(\dfrac{1}{p_3} - M\right)}\left(\frac{-1}{p_3^2}\right) = 0.$$

This gives

$$\frac{p_2}{p_2 - Mp_1} = \frac{p_3}{p_3 - Mp_2} = \frac{1}{1 - Mp_3},$$

and hence $p_2^2 = p_1 p_3$, $p_2 = p_3^2$, and finally

$$p_3 = p_1^{1/3}, \qquad p_2 = p_1^{2/3}.$$

This also gives

$$t_1 = t_2 = t_3 = \frac{T}{\ln 2} \ln \left(\frac{1.15}{2.15/p_1^{1/3} - 1} \right).$$

(For $p_0 = 3.625$, we have $p_1 = 1.686$, $p_2 = 1.417$, $p_3 = 1.190$, $t_1 = 20.482\,\text{min}$, and the total time $3t_1 = 61.447$ min.)

5. For a safe continuous ascent, the external pressure should be the tissue pressure divided by 2.15. The differential equation for $p(t)$ then becomes

$$\frac{dp}{dt} = k(p_e - p) = k\left(\frac{p}{2.15} - p\right) = -k\,\frac{1.15}{2.15}\,p, \qquad k = \ln 2/T$$

$$\frac{dp}{dt} = -.535kp.$$

The solution of this equation is $p = p(0)e^{-.535kt}$, where $p(0)$ is the pressure at time $t = 0$. The diver's depth at time t is related to $p_e(t)(= p(t)/2.15)$ by

$$1 + \frac{d}{33} = p_e(t) = p(0)e^{-.535kt}/2.15.$$

For a long dive at 4 atm and $T = 40$, we have

$$d = 33 \times 1.86 \left(\frac{1}{2}\right)^{.535t/40} - 33$$

$$= 33\left[1.86\left(\frac{1}{2}\right)^{.0134t} - 1\right].$$

The time to ascend to the surface is the value of t at which $d = 0$, that is,

$$t = \frac{1}{.0134}\,\frac{\ln 1.86}{\ln 2} \approx 66.81.$$

6. If the partial pressure of nitrogen is $0.8p$, where p is the tissue pressure, then the maximum safe pressure for the nitrogen is 0.8×2.15, so that the condition $p < 2.15$ is retained. Moreover, if the external gas pressure is p_e, the external nitrogen pressure is $0.8p_e$, and the equation for absorption of nitrogen will be

$$\frac{d}{dt}(.8p) = k(.8p_e - .8p)$$

with initial nitrogen pressure $.8p_0$. Thus, the differential equation for the pressure is the same and the criterion for safe ascent is the same.

7. **Table 5** gives the pressures at the ends of the stops for the half-times 5, 10, 20, 40, and 75 min.

Table 5.

Pressures at the ends of the stops for the dive of **Exercise** 7.

	5	10	20	40	75	Safe pressure at next stop
45 min at 3.58	3.57	3.46	3.04	2.40	1.88	4.08
2 min at 1.9	3.17	3.26	2.96	2.38		3.44
7 min at 1.6		2.62	2.67	2.29		2.795
15 min at 1.3			2.11	2.06		2.15

From the table, we see that a safe pressure has been reached to ascend to the next stop in all cases. The blanks in the 75 column have not been calculated, since they will all be less than 1.88. In the 5 column, the blanks will be less than the corresponding entries in the 10 column, and the final entry in the 10 column will be less than that in the 20 column.

8. First dive at 3.4 atm for 15 min. Pressures will be for $T = 20$, 1.97; for $T = 40$, $p = 1.55$.

   After 60 min at the surface, $p_e = 1$. For $T = 20$, $p = 1.12$; for $T = 40$,

   Descent to 4 atm. Diver may remain until tissue pressure is 2.15. For $T = 20$, this requires 12.77 min; for $T = 40$, 24.12 min.

   The diver must still return to the surface after 12.77 min.

9. First dive 4 atm for 10 min. Pressures: for $T = 20$, $p = 1.88$; for $T = 40$, $p = 1.48$.

   After 60 min at 1 atm: for $T = 20$, $p = 1.11$; for $T = 40$, $p = 1.17$.

   Second dive to 3.4 atm until $p = 2.15$. For $T = 20$, 17.47 min; for $T = 40$, 33.40 min.

   The diver must ascend after 17.47 min.

10. We use **(7)**. The time for ascent is $90/60 = 3/2$ min, so we have

$$p = 1 + \frac{60}{33} \cdot \frac{3}{2} - \frac{60}{33(\ln 2/20)} + \left(3 + \frac{60}{33(\ln 2/20)}\right)\left(\frac{1}{2}\right)^{3/40} = 3.92.$$

A stop of 1.5 min at 10 ft (1.3 atm) reduces the pressure to 3.86 atm.

References

Bachrach, Arthur. 1982. A short history of man in the sea. In Bennett and Elliott [1982], 1–14.

Barnes, Ron. 1987. Compartment models in biology. UMAP Modules in Undergraduate Mathematics and Its Applications: Module 676. *The UMAP Journal* 8 (2): 133–160. Reprinted in *UMAP Modules: Tools for Teaching 1987*, edited by Paul J. Campbell, 207–234. Arlington, MA: COMAP, 1988.

Bennett, Peter B., and David H. Elliott, ed. 1982. *The Physiology and Medicine of Diving.* 3rd ed. London: Bailliere Tindall.

Bornmann, Robert C. 1970. U.S. Navy experiences with decompression from deep helium oxygen saturation excursion diving. In *Human Performance and Scuba Diving*, Proceedings of the Symposium on Underwater Physiology at Scripps Institute of Oceanography, La Jolla, Calif. Chicago, IL: Athletic Institute.

Hammes, Richard B., and Anthony G. Zimos. 1988. *Safe Scuba.* Long Beach, CA: The National Association of SCUBA Diving Schools. Chapter 9.

Hempleman, Henry V. 1982. History of evolution of decompression procedures. In Bennett and Elliott [1982], 319–351.

Hills, B.A. 1977. *Decompression Sickness.* New York: John Wiley & Sons.

Moon, Richard E., Richard D. Vann, and Peter B. Bennett. The physiology of decompression illness. *Scientific American* 273 (2) (August 1995): 70–77.

Vann, Richard D. 1982. Decompression theory and applications. In Bennett and Elliott [1982], 352–382.

Walder, Dennis. 1982. The compressed air environment. In Bennett and Elliott [1982], 15–30.